P9-CCC-377

FOURIER SERIES

G. H. HARDY

W. W. ROGOSINSKI

DOVER PUBLICATIONS, INC.
Mineola, New York

Bibliographical Note

This Dover edition, first published in 1999, is an unabridged republication of the work originally published in 1956 by the Syndics of the Cambridge University Press, London.

Library of Congress Cataloging-in-Publication Data

Hardy, G. H. (Godfrey Harold), 1877–1947.
Fourier series / G.H. Hardy, W.W. Rogosinski.
p. cm.
Originally published: 3rd ed. Cambridge : University Press, 1956.
Includes bibliographical references.
ISBN 0-486-40681-4 (pbk.)
1. Fourier series. I. Rogosinski, Werner, 1894– II. Title.
QA404.H3 1999
515'.2433—dc21 98–48801
CIP

Manufactured in the United States of America
Dover Publications, Inc., 31 East 2nd Street, Mineola, N.Y. 11501

CONTENTS

PREFACE

This tract is based on lectures which each of us has given in Cambridge or elsewhere. There are already a good many books on the subject; but we think that there is still room for one which is written in a modern spirit, concise enough to be included in this series, yet full enough to serve as an introduction to Zygmund's standard treatise.

We have not written for physicists or for beginners, but for mathematicians interested primarily in the theory and with a certain foundation of knowledge. In particular we assume acquaintance with the elements of Lebesgue's theory of integration: it is impossible to understand the theory of Fourier series properly without it, and experience shows that it is well within the powers of any clever undergraduate. The actual knowledge needed here can be acquired quite easily from chapters x–xii of Titchmarsh's *Theory of Functions*. As regards the theory of trigonometrical series, the book is 'officially' complete in itself; but we recognize unofficially that practically all our readers will have some knowledge of the subject (such as the substance of Titchmarsh's chapter xiii) already.

We have naturally been forced to omit much which we should have liked to include. In particular we have no space for the inequalities of Young and Hausdorff, Marcel Riesz's theorem concerning conjugate series, theorems concerning Cesàro summation of general order, or uniqueness theorems involving summable series. And we give no results about special series except a few which we require to illustrate the general theory.

The notes at the end are not systematic; we have inserted only such references and comments as we could make shortly and seemed to us likely to be useful. In particular we make no attempt to give any adequate idea of the history of the subject: Euler, Fourier himself, Poisson and Dirichlet are hardly mentioned. It is quite impossible in an account like this to do any justice to the great mathematicians who founded the theory.

We have to thank Miss S. M. Edmonds, Dr W. H. J. Fuchs, Dr A. J. Macintyre, and Dr A. C. Offord for assistance with the proof-sheets and many valuable criticisms.

G. H. H.
W. W. R.

Cambridge and Aberdeen, Sept. 1943

PREFACE TO SECOND EDITION

G. H. Hardy is no more. British Mathematics has lost its undisputed leader; my Refugee colleagues in this country mourn in him the sincere humanitarian who offered understanding sympathy, advice, and assistance in difficult times; and I myself, if I may claim it, miss a real friend.

No doubt the reader of this Tract will notice in its style the hand of the master: the final draft was written by Hardy. I have not interfered with it in the new edition. A few misprints and mistakes were pointed out to us by various colleagues; the worst are the 'slip' in the proof (ii) of Theorem 31, and the false form of Theorem 59. These are now corrected.

W. W. R.

Newcastle upon Tyne, June 1949

PREFACE TO THIRD EDITION

A few remaining mistakes are now corrected, amongst them the wrong numerical value of the Gibbs constant. An unusual form of Egoroff's theorem has been employed in the proof of Theorem 89. This form is now explicitly stated in § 1.4 and its proof indicated in the Notes.

W. W. R.

Newcastle upon Tyne, September 1955

NOTATIONS

We use the abbreviations t.s. (1), c.s. (3), F.s., F.c. (4), p.p. (5), o.s., n.o.s. (11) in senses defined on the pages indicated in brackets.

The O, o notation is used in the manner now customary: see, e.g., Hardy, *Pure Mathematics* (ed. 8, Cambridge, 1941), p. 164. The symbol $\sim$ is used occasionally (e.g. p. 49) for asymptotic equality, but more generally to express the relation of a function to its F.s.

$\bar{z}$ is the conjugate complex of z. $[x]$ is the integral part of x. Min (x, y) and Max (x, y) are the lesser and greater of x and y. For Max $|f|$ see p. 7.

$\sum_{\alpha}^{\beta} u_n$ denotes a sum over $\alpha \leqq n \leqq \beta$, whether α and β are integral or not. If $\beta < \alpha$, it is 0. We sometimes omit limits in sums and integrals, when it is clear what they are.

$\langle a, b \rangle$ is the closed interval $a \leqq x \leqq b$. We use this symbol only when the closure of the interval is important, using (a, b) when the interval is open or when the distinction is irrelevant.

H is a 'constant', i.e. a number independent of the parameters of the argument, whose precise value is immaterial. We sometimes use ϵ for 'any positive number' (with the emphasis on its possible smallness) without special explanation (e.g. pp. 7, 14).

A new symbol occurring for the first time in a formula without explanation is *defined by* the formula. Thus $C_n(\theta)$ is defined as $c_n e^{ni\theta}$ by (1.1.8), p. 1.

I. GENERALITIES

1.1. Trigonometrical series. A trigonometrical series (t.s.) is a series of the form

$$\tfrac{1}{2}A_0(\theta)+\sum_1^\infty A_n(\theta), \tag{1.1.1}$$

where

$$A_0(\theta)=a_0,\quad A_n(\theta)=a_n\cos n\theta+b_n\sin n\theta \quad (n>0). \tag{1.1.2}$$

We call this series $T(\theta)$ or simply T.

The partial sum of $T(\theta)$, of rank n, is

$$s_n(\theta)=\tfrac{1}{2}A_0(\theta)+\sum_1^n A_m(\theta). \tag{1.1.3}$$

The coefficients a_n and b_n are given, in the first instance, for $n\geqq 0$ and $n\geqq 1$ respectively. We define a_n and b_n, for other integral values of n, by

$$a_{-n}=a_n \;(n>0),\quad b_0=0,\quad b_{-n}=-b_n \;(n>0), \tag{1.1.4}$$

and c_n by

$$c_n=\tfrac{1}{2}(a_n-ib_n); \tag{1.1.5}$$

so that

$$a_n=c_n+c_{-n},\quad b_n=i(c_n-c_{-n}). \tag{1.1.6}$$

Conversely, if the c_n are given, we may define a_n and b_n by (1.1.6). Then

(1.1.7)

$$s_n(\theta)=c_0+\sum_1^n\{(c_m+c_{-m})\cos m\theta+i(c_m-c_{-m})\sin m\theta\}=\sum_{-n}^n c_m e^{mi\theta}.$$

We may therefore also define $T(\theta)$ as

$$\sum_{-\infty}^\infty C_n(\theta)=\sum_{-\infty}^\infty c_n e^{ni\theta}, \tag{1.1.8}$$

and $s_n(\theta)$ by (1.1.7).

We shall call (1.1.1) a 'real' and (1.1.7) a 'complex' t.s. The adjectives refer to the trigonometrical or exponential functions which occur in the series. The coefficients a_n and b_n in (1.1.2) may be complex; but we may suppose them real, when this is convenient, by considering the real and imaginary parts of $T(\theta)$ separately. The series are formal series: there is no implication of their convergence for all or any θ. But (1.1.8) should be thought of as a limiting form of (1.1.7), i.e. as a series in some sense 'equally extended' in the positive and negative directions.

In the simplest cases the series have a sum function $f(\theta)$, and their coefficients can be expressed simply in terms of $f(\theta)$. Suppose, for example, that the series are uniformly convergent. Then, multiplying by $\cos m\theta$ and $\sin m\theta$ or, in the complex case, $e^{-mi\theta}$, integrating term by term over $(-\pi, \pi)$, using the familiar formulae

$$(1.1.9)\qquad \begin{cases} \displaystyle\int_{-\pi}^{\pi} \cos m\theta \cos n\theta \, d\theta = \begin{matrix} 0 & (m \neq n) \\ \pi & (m = n \neq 0), \\ 2\pi & (m = n = 0) \end{matrix} \\ \displaystyle\int_{-\pi}^{\pi} \sin m\theta \sin n\theta \, d\theta = \begin{matrix} 0 & (m \neq n) \\ \pi & (m = n \neq 0), \\ 0 & (m = n = 0) \end{matrix} \\ \displaystyle\int_{-\pi}^{\pi} \cos m\theta \sin n\theta \, d\theta = 0, \\ \displaystyle\int_{-\pi}^{\pi} e^{(n-m)i\theta} \, d\theta = \begin{matrix} 0 & (m \neq n) \\ 2\pi & (m = n) \end{matrix}, \end{cases}$$

and finally replacing m by n, we find that

$$(1.1.10)\qquad \begin{cases} a_n = \dfrac{1}{\pi}\displaystyle\int_{-\pi}^{\pi} f(\theta) \cos n\theta \, d\theta, \quad b_n = \dfrac{1}{\pi}\displaystyle\int_{-\pi}^{\pi} f(\theta) \sin n\theta \, d\theta, \\ c_n = \dfrac{1}{2\pi}\displaystyle\int_{-\pi}^{\pi} f(\theta) \, e^{-ni\theta} \, d\theta. \end{cases}$$

If f is real, a_n and b_n are real, c_n and c_{-n} conjugate. If f is even, $b_n = 0$ and

$$(1.1.11)\qquad a_n = \frac{2}{\pi}\int_{0}^{\pi} f(\theta) \cos n\theta \, d\theta.$$

If f is odd, $a_n = 0$ and b_n may be reduced similarly.

1.2. Trigonometrical series and harmonic functions. It will be useful to begin by indicating the formal connections between the theory of t.s. and the general theory of harmonic and analytic functions, of which in a sense it is a part. In what follows $z = x + iy = re^{i\theta}$ is a complex variable and $\bar{z} = x - iy = re^{-i\theta}$ its conjugate. To fix our ideas, we suppose a_n and b_n *real* (so that

c_n and c_{-n} are conjugate). We also suppose a_n and b_n *bounded* (as they will be in nearly all our work). The power series in r which we write down will then be convergent for $r<1$ and, for a fixed r, uniformly in θ.

If

$$(1.2.1)\quad u(r,\theta) = \sum_{-\infty}^{\infty} c_n r^{|n|} e^{ni\theta} = c_0 + \sum_1^{\infty} c_n r^n e^{ni\theta} + \sum_1^{\infty} c_{-n} r^n e^{-ni\theta},$$

then u is a harmonic function, a solution of either of the equations

$$(1.2.2)\quad \frac{\partial^2 u}{\partial x^2} + \frac{\partial^2 u}{\partial y^2} = 0, \quad \left(r\frac{\partial}{\partial r}\right)^2 u + \frac{\partial^2 u}{\partial \theta^2} = 0.$$

It is real and regular for $r<1$. We can also write

$$(1.2.3)\quad u(r,\theta) = \tfrac{1}{2}a_0 + \sum_1^{\infty} A_n(\theta) r^n,$$

and $T(\theta)$ is the result of writing $r=1$ in either form of u. Now

$$(1.2.4)\quad u(r,\theta) = \tfrac{1}{2}\{F(z) + \overline{F(z)}\},$$

where

$$(1.2.5)\quad F(z) = c_0 + 2\sum_1^{\infty} c_n z^n.$$

Thus u is the real part of $F(z)$. Also, if we write

$$(1.2.6)\quad B_n(\theta) = b_n \cos n\theta - a_n \sin n\theta,$$

we have

$$(1.2.7)\quad F(z) = u(r,\theta) - iv(r,\theta),$$

where

$$(1.2.8)\quad v(r,\theta) = \sum_1^{\infty} B_n(\theta) r^n.$$

We shall call u and v *conjugate harmonic* functions, and the series

$$(1.2.9)\quad \tilde{T}(\theta) = \sum_1^{\infty} B_n(\theta),$$

obtained by writing $r=1$ in (1.2.8), the *conjugate series* (c.s.) of $T(\theta)$.

It will be convenient to prove here two formulae needed in Ch. III. If $C_0 = c_0$, $C_n = 2c_n$ for $n>0$, so that $F(z) = \Sigma C_n z^n$, and $r<1$, then

$$\frac{1}{2\pi}\int_{-\pi}^{\pi} (u-iv)\, e^{-ni\theta}\, d\theta = C_n r^n \quad (n \geqq 0), \quad \frac{1}{2\pi}\int_{-\pi}^{\pi} (u-iv)\, e^{ni\theta}\, d\theta = 0 \quad (n>0).$$

Hence (combining the first equation with the conjugate of the second)

$$(1.2.10)\quad \frac{1}{2\pi}\int_{-\pi}^{\pi} u\,d\theta = \mathbf{R}(C_0),\quad \frac{1}{\pi}\int_{-\pi}^{\pi} ue^{-ni\theta}\,d\theta = \frac{1}{i\pi}\int_{-\pi}^{\pi} ve^{-ni\theta}\,d\theta = C_n r^n$$

for $n > 0$. Actually, our C_0 here is real.

1.3. Trigonometrical Fourier series. Our proof of the formulae (1.1.10) depended on the hypothesis that $T(\theta)$ was uniformly convergent, a drastic assumption unlikely to be satisfied by a t.s. chosen at random. The formulae themselves suggest that we should look at the series from a quite different point of view.

We start from a (real or complex) function $f(\theta)$ integrable (in the sense of Lebesgue) in the interval $(-\pi, \pi)$. It is then convenient to define $f(\theta)$, for all real θ, as a function with period 2π, so that $f(\theta+2\pi) = f(\theta)$, and in particular $f(\pi) = f(-\pi)$, whenever $f(\theta)$ is defined for one of the values of θ in question.

We now define a_n, b_n, and c_n by (1.1.10). We call a_n and b_n the 'real', c_n the complex, *Fourier constants* (F.c.) of $f(\theta)$, and (1.1.1) or (1.1.8) its *Fourier series* (F.s.). We express the fact that a_n and b_n are the F.c. of $f(\theta)$, and (1.1.1) its F.s., by writing $f \sim (a_n, b_n)$ or

$$(1.3.1)\qquad f(\theta) \sim \tfrac{1}{2}a_0 + \sum_{1}^{\infty} (a_n \cos n\theta + b_n \sin n\theta).$$

Similarly we write $f \sim (c_n)$ or $f(\theta) \sim \Sigma c_n e^{ni\theta}$, and call (1.1.1) the 'real', (1.1.8) the 'complex', F.s. of $f(\theta)$. We shall sometimes write $T(f)$ for the F.s. of f, and $\tilde{T}(f)$ for the conjugate t.s.

Since all the functions in (1.1.10) are periodic, we can replace the range of integration by any range $(\xi, \xi+2\pi)$. In particular, it is often convenient to regard $(0, 2\pi)$, rather than $(-\pi, \pi)$, as the fundamental interval.

To say that a t.s. is a F.s. is to say that its coefficients a_n, b_n, or c_n are expressible in the form (1.1.10), that is to say that a certain system of integral equations has a solution. It is plain that the meaning of this statement depends on the definition of integration which we are using. We have adopted Lebesgue's: any restriction or enlargement of the definition of integration would lead to a corresponding change in the class of F.s.

We shall see, for example, that the series

$$(1.3.2)\quad \tfrac{1}{2} + \cos\theta + \cos 2\theta + \dots,\qquad (1.3.3)\quad \frac{\sin 2\theta}{\log 2} + \frac{\sin 3\theta}{\log 3} + \dots$$

are not, in our sense, F.s.; but the coefficients of each series can be expressed in 'Fourier form' by an appropriate generalization of the notion of an integral. Those of (1.3.2) can be expressed in the form

$$\begin{matrix} a_n \\ b_n \end{matrix} = \frac{1}{\pi}\int_{-\pi}^{\pi} \begin{matrix} \cos \\ \sin \end{matrix} n\theta \, d\phi(\theta),$$

a 'Stieltjes integral' in which $\phi(\theta)$ is $-\frac{1}{2}\pi$, 0, and $\frac{1}{2}\pi$ for negative, zero, and positive θ; and those of (1.3.3) in the form (1.1.10), $f(\theta)$ being the sum of the series and the integral for a_n (which is 0) a 'principal value' in the sense of Cauchy.

A t.s. may or may not converge, and it may or may not be a F.s.; and there is no obvious correlation between the two properties (though the simplest series may be expected to have both). The series (1.3.3) converges for all θ, but is not a F.s.; on the other hand there are F.s. which do not converge for any θ. It is not even plain *a priori* that a t.s., known to converge and to be a F.s., is necessarily the F.s. of its sum.

Trigonometrical series are a special class of *orthogonal* series; and there is a considerable part of their theory which is best regarded as part of the theory of these more general series, and which we shall develop from this point of view in Ch. II. But we must begin by a short résumé of certain parts of the theory of functions of a real variable with which we shall assume the reader to be acquainted.

1.4. Measure and integration. We take as known the elements of Lebesgue's theory of measure and integration. We denote by $L(a, b)$, or simply L, the class of functions $f(x)$ integrable, in Lebesgue's sense, over (a, b). The interval of integration will always be finite. We shall sometimes say 'f is L' for 'f belongs to L'. We regard the integral of a *non-negative* function f as defined whenever f is measurable, and having a finite or infinite value according as f is or is not L.

We call a set of measure 0 a *null set*: null sets are irrelevant in the theory of integration. If f and g differ only in a null set, we say that they are *equivalent*, and write $f \equiv g$. We also say that $f = g$ for almost all x, or almost always (or almost everywhere), or 'p.p.' (*presque partout*). If $f \equiv 0$, we say that f is *null*. We write mE for the measure of a set E.

We shall sometimes use letters for other classes of functions besides L; in particular we shall denote by B, C, C_k, and V the classes of bounded functions, continuous functions, functions with k continuous derivatives, and functions of bounded variation*.

We take for granted the classical theorems concerning integration and differentiation, the theorems of partial integration and substitution, the first and second mean value theorems, and the two most familiar theorems concerning passage to the limit under the integral sign. These last are: (i) if $f_n(x) \to f(x)$ p.p. and $|f_n(x)| \leqq \phi(x)$, where $\phi(x)$ is L and independent of n, then

$$\int f_n(x)\,dx \to \int f(x)\,dx; \tag{1.4.1}$$

(ii) the conclusion is also true if $f_n(x)$ increases with n for all, or almost all, x provided that $\int f_n(x)\,dx \neq -\infty$. In case (i) we shall say that $f_n(x)$ converges *dominatedly* to $f(x)$. In particular the conditions are satisfied if $f_n(x) \to f(x)$ for all x, and $|f_n(x)| \leqq H$, in which case we shall say that $f_n(x)$ converges *boundedly* to $f(x)$. In case (ii) it is to be understood that the limit function $f(x)$, which certainly exists p.p., may be infinite for some x, and that $\int f(x)\,dx$ may also be infinite, in which case it is to be replaced by ∞ in (1.4.1). Finally, the integrations in (1.4.1) may be taken over the whole interval (a, b) or any measurable set contained in it. A useful addition to (i) is 'Fatou's lemma': if $f_n(x) \geqq 0$ and $f_n(x) \to f(x)$ p.p., then

$$\int f(x)\,dx \leqq \underline{\lim} \int f_n(x)\,dx.$$

We require one theorem concerning the inversion of the order of integration, Fubini's theorem that

$$\int dx \int f\,dy = \int dy \int f\,dx = \iint f\,dx\,dy$$

whenever $f(x, y)$ is integrable. The integrals may be infinite when $f \geqq 0$.

We shall occasionally use the notion of the Stieltjes integral of a continuous function with respect to a function of V, the rule for the integration of such an integral by parts, and the theorem that $\left|\int f\,d\phi\right| \leqq \mathrm{MV}$, where M is the maximum of $|f|$ and V the total variation of Φ. In Ch. VI we use Egoroff's theorem in two forms: (i) if $f_n(x) \to f(x)$ p.p. in E, then $f_n(x) \to f(x)$ uniformly in a set E^*

* A complex function is V when its real and imaginary parts are V.

included in E and of measure greater than $mE-\epsilon$; (ii) a similar conclusion holds if $f_h(x)\to f(x)$ as $h\to 0$ p.p. in E provided that each $f_h(x)$ is continuous in E.

1.5. The classes L^p. We shall say that f is L^p if f is measurable and $|f|^p$ is L: we shall always suppose that $p\geqq 1$. When p is 1, L^p is L. If f is L^q, and $1\leqq p<q$, then f is L^p.

We shall write

(1.5.1)
$$\mathbf{N}_p(f)=||f||_p=\left(\int_a^b|f|^p\,dx\right)^{1/p},\quad \mathbf{M}_p(f)=\left(\frac{1}{b-a}\int_a^b|f|^p\,dx\right)^{1/p}.$$

If f is not L^p, $\mathbf{N}_p(f)$ and $\mathbf{M}_p(f)$ are infinite. We call $\mathbf{N}_p(f)$ the *norm*, $\mathbf{M}_p(f)$ the *mean*, of f, for the interval (a,b) and index p. They differ only by a factor $(b-a)^{1/p}$; but this difference is important.

If $p>1$, we define p' by

(1.5.2)
$$p'=\frac{p}{p-1},\quad \frac{1}{p}+\frac{1}{p'}=1.$$

Then $p'>1$: if $p<2$, $p'>2$. We call p and p' *conjugate* indices, L^p and $L^{p'}$ conjugate classes. The class L^2 is self-conjugate. If $p=1$, we interpret p' as ∞, and conversely. We shall define the class L^∞, conjugate to L, in a moment.

The means $\mathbf{M}_p(f)$ have three fundamental properties. The first, viz.

(1.5.3)
$$\mathbf{M}_1(fg)\leqq \mathbf{M}_p(f)\,\mathbf{M}_{p'}(g),$$

is *Hölder's inequality* (Schwarz's inequality when $p=2$). The second, viz.

(1.5.4)
$$\mathbf{M}_p(f+g)\leqq \mathbf{M}_p(f)+\mathbf{M}_p(g),$$

is *Minkowski's inequality*. The third, viz.

(1.5.5)
$$\mathbf{M}_q(f)\leqq \mathbf{M}_p(f)\quad (q<p),$$

states that $\mathbf{M}_p(f)$ is, for given f, an increasing function of p. The norms $\mathbf{N}_p(f)$ have the first two properties but not the third.

When $p\to\infty$,

(1.5.6)
$$\mathbf{N}_p(f)\to \mathrm{Max}\,|f|,\quad \mathbf{M}_p(f)\to \mathrm{Max}\,|f|,$$

where $\mathrm{Max}\,|f|$ is the 'effective upper bound' of $|f|$, i.e. the smallest η such that $|f|\leqq\eta$ p.p. It is natural to define L^∞ as the class of f for which $\text{Max}\,|f|$ is finite. This is the class of 'effectively bounded' functions, or functions equivalent to bounded functions. We may now write

$$\mathbf{N}_\infty(f) = \mathbf{M}_\infty(f) = \text{Max}\,|f|; \tag{1.5.7}$$

and it may be verified at once that (1.5.3)–(1.5.5) remain true when p or p' is infinite.

1.6. Space L^p and its metric. The theory of the classes L^p, and of the inequalities associated with them, is much illuminated by the use of geometrical language. The class L^p defines a *functional space*, each function defining a *point* of the space. We do not distinguish two functions equivalent to one another, so that each point is a class of equivalent functions. In particular the *origin* 0 is the class of null functions. The space L^2, which is particularly important, is called *Hilbert space*. We define the *distance* $\delta_p(f, g)$ between f and g, in L^p, by

$$\delta_p(f, g) = \mathbf{N}_p(f-g), \tag{1.6.1}$$

or $\delta(f, g) = \mathbf{N}(f-g)$, omitting the suffix when it is plain of what space we are speaking. In particular, $\mathbf{N}(f)$ is the distance of f from the origin. If $p = \infty$, then

$$\delta(f, g) = \text{Max}\,|f-g|. \tag{1.6.2}$$

We can also define space C, the space of all *continuous* functions: here also distance is defined by (1.6.2), but 'Max' is the ordinary maximum.

If we take $f = f_1 - f_2$, $g = f_2 - f_3$ in (1.5.4), it becomes

$$\delta(f_1, f_3) \leqq \delta(f_1, f_2) + \delta(f_2, f_3), \tag{1.6.3}$$

and appears as an extension of the theorem that a side of a triangle cannot exceed the sum of the other two.

We can now set up a metric in space L^p (or C), and carry over to it the ideas of the ordinary theory of sets of points. One such idea is particularly important for our purposes, that of a class of functions *dense* in a wider class. Suppose that S_1 is a sub-class of L^p and S_2 a sub-class of S_1. Then we say that S_2 is *dense* (L^p) *in* S_1 (or simply dense in S_1) if, given any ϕ of S_1 and any positive ϵ, there is a ψ of S_2 such that $\delta_p(\phi, \psi) < \epsilon$. It follows from (1.6.3) that the relation of density is transitive: if S_2 is dense in S_1, and S_3 in S_2, then S_3 is dense in S_1. In such statements, of course, a fixed metric is presupposed.

It also follows from (1.5.5) that if S_2 is dense (L^p) in S_1, and $1 \leqq q < p$, then S_2 is dense (L^q) in S_1.*

There is one proposition concerning density which we shall use so often that we state it as a formal theorem.

Theorem 1. *If $1 \leqq p < \infty$, then the classes L^q ($q > p$), L^∞, B, C, and C_k are dense in L^p.*

The theorem remains true if all the functions are restricted by periodicity. We shall prove later that the class of algebraical polynomials is dense in L^p, and the class of trigonometrical polynomials dense in the class of periodic functions of L^p.

1.7. Convergence in L^p (*strong convergence*). If f_n and f are L^p, and

$$\delta_p(f_n, f) \to 0 \tag{1.7.1}$$

when $n \to \infty$, or (what is the same thing) if $\mathbf{N}_p(f_n - f)$ tends to 0, then we say that f_n *tends* (L^p) *to* f, and write

$$f_n \to f(L^p). \tag{1.7.2}$$

We shall also say that f_n *tends strongly to f with index p* (omitting the reference to the index if there is no ambiguity). When $p = \infty$, $\delta(f_n, f) = \operatorname{Max} |f_n - f|$, and strong convergence is 'uniform convergence p.p.': $f_n \to f\,(L^\infty)$ if $f_n \equiv f_n^*$ and $f_n^* \to f$ uniformly.

A strong limit is 'effectively unique': if $f_n \to f\,(L^p)$ and $f_n \to g\,(L^p)$, then $f \equiv g$.

If $f_n \to f\,(L^p)$ and $1 \leqq q < p$, then $f_n \to f\,(L^q)$.

If $f_n \to f\,(L^p)$, then $\mathbf{N}_p(f_n) \to \mathbf{N}_p(f)$.

If $f_n \to f\,(L^p)$ and $g_n \to g\,(L^{p'})$, then $f_n g_n \to fg\,(L)$ and

$$\int f_n g_n \, dx \to \int fg \, dx. \tag{1.7.3}$$

In particular this is true if g is $L^{p'}$ and $g_n = g$ for all n.

If $f_n \to f$ p.p., $|f_n| \leqq \phi$, where ϕ is L^p and independent of n, and $1 \leqq p < \infty$, then $f_n \to f\,(L^p)$.

* It is not true that $\mathbf{N}_q(\phi - \psi) \leqq \mathbf{N}_p(\phi - \psi)$, but (owing to the arbitrariness of ϵ) the powers of $b - a$ involved in (1.5.5) do not affect the conclusion.

The fundamental theorem concerning strong convergence is

Theorem 2. *In order that f_n should converge strongly, with index p, to a function f of L^p, it is necessary and sufficient that*

$$\int |f_m - f_n|^p \, dx \to 0$$

when m and n tend to infinity. There is then a sub-sequence (n_k) such that $f_{n_k} \to f$ for almost all x.*

Theorem 2 is the analogue, for space L^p, of Cauchy's theorem concerning ordinary limits. When $p = \infty$ it reduces (apart from the neglect of null sets) to the corresponding theorem about uniform convergence. Strong convergence does not imply convergence p.p. (or for any x), nor is the converse true. But it follows from Theorem 2 that, if $f_n \to f \, (L^p)$ and $f_n \to f^*$ p.p., then $f \equiv f^*$.

Finally there is one theorem which we shall often use.

Theorem 3. *If $1 \leqq p < \infty$ and f is L^p, then*

$$\int_a^b |f(x+h) - f(x)|^p \, dx \to 0$$

when $h \to 0$, i.e. $f(x+h) \to f(x) \, (L^p)$.

The integral involves the values of f for certain values of x outside (a, b). We may either suppose these values to be 0, or regard $f(x)$ as a function with period $b - a$.

1.8. The resultant of two periodic functions. The *resultant* (*Faltung*) of two functions f and g, each with period $b - a$, over (a, b), is

$$r(x) = \frac{1}{b-a} \int_a^b f(x-y) \, g(y) \, dy. \tag{1.8.1}$$

The resultant is also periodic, and symmetrical in f, g. The fundamental properties of $r(x)$ are as follows.

Theorem 4. *If f and g are L, then r is L (and so finite for almost all x). Also*

$$\frac{1}{b-a} \int_a^b r \, dx = \frac{1}{b-a} \int_a^b f \, dx \, \frac{1}{b-a} \int_a^b g \, dx, \tag{1.8.2}$$

and

$$\mathbf{M}_1(r) \leqq \mathbf{M}_1(f) \, \mathbf{M}_1(g). \tag{1.8.3}$$

* I.e. that $\int |f_m - f_n|^p \, dx < \epsilon$ for every positive ϵ and $m \geqq M(\epsilon)$, $n \geqq N(\epsilon)$.

For $f(x-y)$ and therefore $f(x-y)\,g(y)$ are measurable functions of (x,y), and

$$\int_a^b dx \int_a^b f(x-y)\,g(y)\,dy = \int_a^b g(y)\,dy \int_a^b f(x-y)\,dx = \int_a^b g(y)\,dy \int_a^b f(x)\,dx,$$

$$\int_a^b dx \left| \int_a^b f(x-y)\,g(y)\,dy \right| \leqq \int_a^b |\,g(y)\,|\,dy \int_a^b |\,f(x-y)\,|\,dx$$

$$= \int_a^b |\,g(y)\,|\,dy \int_a^b |\,f(x)\,|\,dx,$$

the inversions being justified by Fubini's theorem.

Theorem 5. *If f is L^p and g is $L^{p'}$, then r is continuous.*

Since r is symmetrical in f and g, we may suppose $p < \infty$. Then

$$|\,r(x+h)-r(x)\,| = \left| \frac{1}{b-a} \int_a^b \{f(x+h-y)-f(x-y)\}\,g(y)\,dy \right|$$

$$\leqq \mathbf{M}_p\{f(x+h)-f(x)\}\,\mathbf{M}_{p'}(g);$$

and the first factor tends to 0, by Theorem 3.

1.9. Orthogonal systems in L^2. In our applications of the ideas of §§ 1.5–1.8, p will usually be 1, 2, or ∞ (though we shall state and prove theorems for general p when doing so does not introduce any material complication). In this section it will be essential that the functions ϕ_n considered should be L^2.

Suppose that (ϕ_n), for $n = 0, 1, 2, \ldots$, is a system of non-null functions of $L^2(a,b)$. If

$$(1.9.1) \qquad (\phi_m, \phi_n) = \int_a^b \phi_m \bar{\phi}_n\,dx = 0$$

whenever $m \neq n$, we say that (ϕ_n) is an *orthogonal system* (o.s.) in (a, b). If also

$$(1.9.2) \qquad (\phi_n, \phi_n) = \int_a^b |\,\phi_n\,|^2\,dx = \|\,\phi_n\,\|_2^2 = 1$$

for every n, then we say that (ϕ_n) is a *normal orthogonal system* (n.o.s.) in (a, b).

If f also is L^2, and

$$(1.9.3) \quad c_n = \frac{1}{||\phi_n||^2}\int_a^b f\bar{\phi}_n\,dx = \frac{(f,\phi_n)}{(\phi_n,\phi_n)} \quad (n = 0, 1, 2, \ldots),$$

then we call the c_n the *Fourier constants* (F.c.) of f with respect to (ϕ_n), and $\Sigma c_n \phi_n$ its *Fourier series* (F.s.); and we write $f \sim (c_n)$ or

$$(1.9.4) \qquad f \sim \Sigma c_n \phi_n.$$

All these definitions are similar to those of § 1.3. They are stated only for functions of L^2, since (f, ϕ_n) will not usually exist for other f; but they may be applied more widely if the ϕ_n are subject to further restrictions. If, for example, ϕ_n is bounded for each n, then any f of L has a F.s. defined as in (1.9.3) and (1.9.4). The trigonometrical systems with which we are primarily concerned satisfy this condition.

Geometrically, we may regard ϕ_n as a point, or as a vector (of position measured from 0) in space L^2. If (ϕ_n) is orthogonal, the vectors are orthogonal; if (ϕ_n) is normal, they are unit vectors. If (ϕ_n) is a n.o.s., f is L^2, and $f \sim \Sigma c_n \phi_n$, then the c_n may be regarded as 'coordinates' of f with respect to a system of 'orthogonal axes' corresponding to the ϕ_n.

A *polynomial* in the ϕ_n is a finite linear combination

$$P(\phi) = a_0\phi_0 + a_1\phi_1 + \ldots + a_n\phi_n$$

of the ϕ_n, with constant coefficients. If no $P(\phi)$ is null unless all its coefficients vanish, then we say that the ϕ_n are *linearly independent*: it is only in this case that they form an appropriate basis for a coordinate system. In particular, no member of a linearly independent system is null.

These last definitions are independent of orthogonality. There is a process of 'orthogonalization' by which we can derive an o.s. (ϕ_n) from any linearly independent system (ψ_n) of functions of L^2. Each ϕ_n is a polynomial in the ψ_n (involving $\psi_0, \psi_1, \ldots, \psi_n$ only), the ϕ_n also are linearly independent, and each ψ_n is a polynomial in the ϕ_n. When we have formed the ϕ_n, we can normalize them by multiplication by appropriate factors. The process is an analogue for space L^2 of transformation from oblique to rectangular axes in ordinary geometry.

1.10. Examples of orthogonal systems. (1) The system

$$(E) \qquad e^{nix} \quad (-\infty < n < \infty)^{\star}$$

is orthogonal in $(-\pi, \pi)$, or any interval of length 2π. The system

$$(E') \qquad (2\pi)^{-\frac{1}{2}} e^{nix} \quad (-\infty < n < \infty)$$

is a n.o.s.

(2) The system

$$(T) \qquad \tfrac{1}{2}, \cos x, \sin x, \cos 2x, \sin 2x, \ldots$$

is orthogonal in $(-\pi, \pi)$. The system

$$(T') \qquad (2\pi)^{-\frac{1}{2}}, \pi^{-\frac{1}{2}}\cos x, \pi^{-\frac{1}{2}}\sin x, \pi^{-\frac{1}{2}}\cos 2x, \ldots$$

is a n.o.s.

(3) The systems

$$(C) \qquad \tfrac{1}{2}, \cos x, \cos 2x, \ldots, \qquad (S) \quad \sin x, \sin 2x, \ldots$$

are orthogonal in $(0, \pi)$.

(4) The system of Legendre polynomials

$$P_n(x) = \frac{1}{2^n n!}\left(\frac{d}{dx}\right)^n (x^2-1)^n = \frac{2n!}{2^n(n!)^2}(x^n - \ldots) \quad (n = 0, 1, 2, \ldots)$$

is orthogonal in $(-1, 1)$. It becomes a n.o.s. when multiplied by $(n+\frac{1}{2})^{\frac{1}{2}}$. It is the system obtained by orthogonalizing the system (x^n) by the process referred to in § 1.9.

1.11. Some further generalities. If (ϕ_n) is an o.s., then any series $\Sigma c_n \phi_n$ is a *series of orthogonal functions* or *orthogonal series*. It will not necessarily be a F.s., in the sense of § 1.9; it will not necessarily be convergent, or 'summable'; nor, if it is convergent or summable, will it necessarily be the F.s. of its sum. These are just the problems which we have to consider for t.s.

In the next chapter we shall prove the most fundamental theorems concerning orthogonal series in L^2. There are a few more generalities to be inserted here.

A system (ψ_n), not necessarily orthogonal, is said to be *complete in L^p*, where $1 \leqq p \leqq \infty$, or in C, if there is no non-null function of L^p (or C) orthogonal to every ψ_n, i.e. if

$$\int_a^b f\bar{\psi}_n\, dx = 0 \quad (n = 0, 1, 2, \ldots)$$

$\star$ It is convenient in this case to number the ϕ_n in this way instead of from 0 to ∞.

implies $f \equiv 0$. The definition is useful only if the ψ_n belong to the conjugate class $L^{p'}$ (or L). It is plain from the definition that, if (ψ_n) is complete in L^p, it is complete in L^q for $q > p$, and in C.

A system (ψ_n) of functions of L^p (or C) is said to be *closed in* L^p (or C) if the system of polynomials $P(\psi)$ is dense in L^p (or C). If (ψ_n) is closed in L^p, and $1 \leqq q < p$, then (ψ_n) is closed in L^q. For if f is any function of L^q there is, by Theorem 1, a g of L^p such that $\delta_q(f-g) < \frac{1}{2}\epsilon$; and, by hypothesis, there is a polynomial Ψ for which $\delta_p(g, \Psi) < \frac{1}{2}\epsilon(b-a)^{1/p-1/q}$ so that $\delta_q(g, \Psi) < \frac{1}{2}\epsilon$, by (1.5.5). If (ϕ_n) is closed in C, then it is closed in L^p for $1 \leqq p < \infty$ (but not necessarily in L^∞). We shall see in Ch. II that the system (x^n) is complete in L and closed in C; incidentally that it is both complete and closed in L^p when $1 \leqq p < \infty$.

If the system (ψ_n) of § 1.9 is complete (or closed) in L^p, and (ϕ_n) is the orthogonalized system, then (ϕ_n) is complete (or closed). The process of orthogonalization, however, presupposes that every ψ_n is L^2.

If (ϕ_n) is a complete n.o.s., and f and g have the same F.c. with respect to (ϕ_n), then $f \equiv g$.

We end this chapter with three useful, though almost trivial, theorems. In these we suppose, to avoid minor complications, that (ϕ_n) is a n.o.s. of functions of L^∞.

Theorem 6. *If*

$$\Phi_n = \sum_0^n c_m \phi_m \to f \ (L^p),$$

then $\Sigma c_n \phi_n$ is the Fourier series of f.

For, by (1.7.3),

$$\int f\bar{\phi}_m \, dx = \lim_{n\to\infty} \int \Phi_n \bar{\phi}_m \, dx = \lim_{n\to\infty} c_m = c_m.$$

Theorem 7. *If $\Sigma c_n \phi_n$ is dominatedly convergent (in particular, if it is boundedly or uniformly convergent), then it is the Fourier series of its sum.*

For we may multiply by $\bar{\phi}_m$ and integrate term by term.

Theorem 8. *If (ϕ_n) is complete, and the Fourier series of f converges dominatedly to s, then $s \equiv f$.*

For it is the F.s. of its sum s, by Theorem 7. Hence s and f have the same F.c., and therefore $s \equiv f$. In particular, $s = f$ at any point where both functions are continuous.

II. FOURIER SERIES IN HILBERT SPACE

2.1. General Fourier series in L^2. In what follows (ϕ_n) is a n.o.s. of functions of $L^2(a, b)$, f is any function of L^2, and $\Sigma c_n \phi_n$ is its F.s. We call

$$f_n = \sum_0^n c_m \phi_m \tag{2.1.1}$$

the *Fourier polynomial*, of rank n, of f. $\mathbf{N}$ is $\mathbf{N}_2$.

Theorem 9. *If*

$$\Phi_n = \sum_0^n \gamma_m \phi_m \tag{2.1.2}$$

is any polynomial in the ϕ, *then*

$$\mathbf{N}^2(f-\Phi_n) = \mathbf{N}^2(f) - \sum_0^n |c_m|^2 + \sum_0^n |c_m - \gamma_m|^2. \tag{2.1.3}$$

In particular

$$\mathbf{N}^2(f-f_n) = \mathbf{N}^2(f) - \sum_0^n |c_m|^2. \tag{2.1.4}$$

It is plain by term-by-term integration that

$$\int_a^b f\bar{\Phi}_n\,dx = \sum_0^n c_m\bar{\gamma}_m, \quad \int_a^b |\Phi_n|^2\,dx = \sum_0^n |\gamma_m|^2.$$

Hence

$$\begin{aligned}
\mathbf{N}^2(f-\Phi_n) &= \int_a^b (f-\Phi_n)(\bar{f}-\bar{\Phi}_n)\,dx \\
&= \mathbf{N}^2(f) - \sum_0^n c_m\bar{\gamma}_m - \sum_0^n \bar{c}_m\gamma_m + \sum_0^n |\gamma_m|^2 \\
&= \mathbf{N}^2(f) - \sum_0^n |c_m|^2 + \sum_0^n (c_m-\gamma_m)(\bar{c}_m-\bar{\gamma}_m),
\end{aligned}$$

which is (2.1.3). As corollaries, we have

Theorem 10. *Among all polynomials Φ_n of given degree n, that which gives the best mean square approximation to f is the Fourier polynomial f_n.*

Theorem 11. *If $f \sim (c_n)$, then*

$$\sum_0^\infty |c_n|^2 \leq \mathbf{N}^2(f) = \int_a^b |f|^2\,dx. \tag{2.1.5}$$

2.2. The Riesz-Fischer theorem. Our next theorem lies deeper, since it depends on Theorem 2 (for $p = 2$).

Theorem 12. *Suppose that* $\Sigma |c_n|^2 < \infty$. *Then there is a function* f *of* L^2 *whose Fourier constants are the* c_n, *and such that* $f_n \to f\,(L^2)$, *i.e.*

$$\int_a^b |f_n - f|^2\,dx \to 0, \tag{2.2.1}$$

and

$$\int_a^b |f|^2\,dx = \sum_0^\infty |c_n|^2. \tag{2.2.2}$$

These results are almost immediate corollaries of Theorem 2. If $n > m$, then

$$\int_a^b |f_m - f_n|^2\,dx = \sum_{m+1}^{n} |c_k|^2 \to 0$$

when m and n tend to infinity. Hence there is an f of L^2 satisfying (2.2.1). By Theorem 6, with $p = 2$, the c_n are the F.c. of f. Finally, after (2.1.4), (2.2.1) is equivalent to (2.2.2).

2.3. Complete systems and Parseval's theorem. So far we have made no use of the notion of completeness: we now suppose that (ϕ_n) is complete. It then follows, first that the f of Theorem 12 is effectively unique, and secondly that (2.2.1) and (2.2.2) are true for any f of L^2 whose F.c. are the c_n.

Theorem 13. *If* (ϕ_n) *is a complete normal orthogonal system,* f *is* L^2, *and* $f \sim (c_n)$, *then* f *and the* c_n *satisfy* (2.2.1) *and* (2.2.2).

For $\Sigma |c_n|^2 < \infty$, by Theorem 11, and so f is equivalent to the f of Theorem 12.

Theorem 14. *If* f *and* F *are* L^2, $f \sim (c_n)$, *and* $F \sim (C_n)$, *then*

$$\int_a^b f\bar{F}\,dx = \sum_0^\infty c_n \bar{C}_n. \tag{2.3.1}$$

The series is absolutely convergent.

For
$$\int_a^b f_n \bar{F}\,dx = \sum_0^n c_m \int_a^b \phi_m \bar{F}\,dx = \sum_0^n c_m \bar{C}_m.$$

But $f_n \to f\,(L^2)$, by Theorem 13, and so $\int f_n\bar{F}\,dx \to \int f\bar{F}\,dx$, by (1.7.3). The series is absolutely convergent because $\Sigma |c_n|^2$ and $\Sigma |C_n|^2$ are convergent.

The name 'Parseval's theorem' is sometimes applied to (2.2.2), sometimes to the more general equation (2.3.1). It will be observed that the theorem appears here as a rather trivial corollary of the Riesz-Fischer theorem. It is not trivial for any particular system; but its difficulties lie in the proof of the completeness of the system, here taken as a hypothesis.

2.4. Mercer's theorem. It follows from Theorem 11 that the F.c. of a function of L^2 tend to zero. If we suppose (ϕ_n) subject to further conditions, then (as we remarked in § 1.9) we can define the F.c. of wider classes of functions. A particularly important case (including the trigonometrical cases) is that in which the ϕ_n are uniformly bounded.

Theorem 15. *If (ϕ_n) is a normal orthogonal system, and $|\phi_n| \leqq H$, then the Fourier constants of any f of L tend to zero.*

By Theorem 1, there is an F of L^2 such that $\int |f-F|\,dx < \epsilon$. If $F \sim (C_n)$, then

$$|c_n| = \left|\int f\bar{\phi}_n\,dx\right| \leqq \left|\int F\bar{\phi}_n\,dx\right| + \int |f-F|\,|\phi_n|\,dx \leqq |C_n| + H\epsilon.$$

But $C_n \to 0$, since F is L^2; and therefore $|c_n| < 2H\epsilon$ for sufficiently large n.

2.5. Closure and completeness. It results from our analysis that closure and completeness are equivalent in L^2 (whether the system of functions considered is orthogonal or not).

Theorem 16. *A system of functions of $L^2(a, b)$ is closed if and only if it is complete.*

It is sufficient, after § 1.9, to prove the equivalence for the system (ϕ_n) obtained by orthogonalizing and normalizing the given system (ψ_n).

(i) If (ϕ_n) is complete, and f is L^2, then $f_n \to f(L^2)$, by Theorem 13. Since f_n is a polynomial Φ_n, (ϕ_n) is closed.

(ii) Suppose that (ϕ_n) is closed, that f is L^2, and that all the F.c. of f are 0. Since (ϕ_n) is closed, there is a sequence of polynomials Φ_n for which $\Phi_n \to f(L^2)$, i.e. $\mathbf{N}(f-\Phi_n) \to 0$; and therefore, by Theorem 10, $\mathbf{N}(f-f_n) \to 0$. But $f_n = 0$, and so $\mathbf{N}f = 0$, $f \equiv 0$. Hence (ϕ_n) is complete.

It is natural to ask what is true in other spaces L^p, where there is no such equivalence. The answer is given by the two theorems which follow. We prove Theorem 17, but the proof of Theorem 18 depends on a concept, that of 'weak convergence', which we shall not use, and we therefore omit it.

Theorem 17. *If $1 \leqq p \leqq \infty$, and (ψ_n) is closed in L^p, then it is complete in $L^{p'}$.*

Theorem 18. *If $1 < p \leqq \infty$, and (ψ_n) is complete in L^p, then it is closed in $L^{p'}$.*

To prove Theorem 17, suppose that (ψ_n) is closed in L^p, that f is $L^{p'}$, and that $\int f\psi_n \, dx = 0$ for every n. If we write $f = |f| e^{i\vartheta}$, then $e^{-i\vartheta} = e^{-i\vartheta(x)}$ is measurable and bounded and so belongs to L^p. There is therefore a sequence of polynomials Ψ_n such that $\Psi_n \rightarrow e^{-i\vartheta}$ (L^p); and

$$\int |f| \, dx = \int fe^{-i\vartheta} \, dx = \lim \int f\Psi_n \, dx = 0,$$

by (1.7.3), so that $f \equiv 0$. Hence (ψ_n) is complete in $L^{p'}$.

Theorem 18 is not true in the case excluded $(p = 1)$.

2.6. Completeness of the trigonometrical systems. We shall now prove that the trigonometrical systems (E) and (T) are complete in L (*a fortiori* in L^2).

Theorem 19. *The trigonometrical systems (E) and (T) are complete in $L(-\pi, \pi)$: if f is $L(-\pi, \pi)$, and the Fourier constants of f are all 0, then f is null.*

It is plainly indifferent which system we select: we take (T), and we begin by proving that (T), if complete in C, is complete in L. Suppose then that f is a real function of L and that

$$F(x) = \int_0^x f(y) \, dy - \tfrac{1}{2}a_0 x,$$

so that F is continuous and periodic. If $F \sim (A_n, B_n)$, and $n \geqq 1$, then

$$A_n = \frac{1}{\pi}\int_{-\pi}^{\pi} F(x) \cos nx \, dx = -\frac{1}{n\pi}\int_{-\pi}^{\pi} f(x) \sin nx \, dx = -\frac{b_n}{n},$$

by partial integration; and similarly $B_n = a_n/n$.* If a_n and b_n are 0 for all n, A_n and B_n are 0 for $n \geqq 1$, and all the F.c. of $F - \frac{1}{2}A_0$ are 0. Hence, if (T) is complete in C, $F - \frac{1}{2}A_0 \equiv 0$. Since F is continuous, $F = \frac{1}{2}A_0$, $f \equiv \frac{1}{2}a_0 = 0$.

It is therefore sufficient to prove (T) complete in C. Let us assume provisionally that, given any positive δ and η, there is a trigonometrical polynomial $T_n(x)$ such that

$$(2.6.1) \quad T_n(x) \geqq 0, \quad \int_{-\pi}^{\pi} T_n(x)\,dx = 1, \quad T_n(x) \leqq \eta \quad (\delta \leqq |x| \leqq \pi).$$

Suppose now that f is continuous and not null, that its maximum modulus is M, and that all of a_n and b_n are 0. Then

$$\int_{-\pi}^{\pi} f(x+\xi)\,T_n(x)\,dx = \int_{\xi-\pi}^{\xi+\pi} f(x)\,T_n(x-\xi)\,dx = 0$$

for every ξ. Since $f \not\equiv 0$, there are a $c \neq 0$ and a ξ such that $f(\xi) = c$: we may suppose c positive. Since f is continuous, there is a positive δ such that $f > \frac{1}{2}c$ throughout $(\xi - \delta, \xi + \delta)$; and then

$$0 = \int_{-\pi}^{\pi} f(x+\xi)\,T_n(x)\,dx \geqq \tfrac{1}{2}c \int_{-\delta}^{\delta} T_n(x)\,dx - \mathrm{M}\left(\int_{-\pi}^{-\delta} + \int_{\delta}^{\pi}\right) T_n(x)\,dx$$

$$= \tfrac{1}{2}c - (\tfrac{1}{2}c + \mathrm{M})\left(\int_{-\pi}^{-\delta} + \int_{\delta}^{\pi}\right) T_n(x)\,dx \geqq \tfrac{1}{2}c - (\tfrac{1}{2}c + \mathrm{M})\,2\pi\eta.$$

This is a contradiction if η is sufficiently small.

We have still to show that there is a T_n satisfying (2.6.1). We may, for example, take

$$T_n = \frac{(1+\cos x)^n}{\displaystyle\int_{-\pi}^{\pi} (1+\cos x)^n\,dx} = \frac{(\cos \frac{1}{2}x)^{2n}}{\displaystyle\int_{-\pi}^{\pi} (\cos \frac{1}{2}x)^{2n}\,dx}.$$

It is obvious that this T_n satisfies the first two conditions (2.6.1); and

$$T_n < \frac{(\cos \frac{1}{2}\delta)^{2n}}{\displaystyle\int_0^{\frac{1}{2}\delta} (\cos \frac{1}{2}x)^{2n}\,dx} < \frac{2}{\delta}\left(\frac{\cos \frac{1}{2}\delta}{\cos \frac{1}{4}\delta}\right)^{2n} \to 0$$

if $\delta \leqq x \leqq \pi$ and $n \to \infty$, so that it also satisfies the third for sufficiently large n.

2.7. The Parseval and Riesz-Fischer theorems for trigonometrical series. We now restate the substance of Theorems 10–14 in the form which it takes for the systems (E) and (T). We

* For the value of A_0 see Theorem 43 (§ 3.8).

state the results for (T) only in the most important case, in which the functions are real.

Theorem 20. *Among all complex trigonometrical polynomials of degree n, that which gives the best mean square approximation to a given f of L^2 is the Fourier polynomial f_n of f.*

If f is L^2 and $f \sim (c_n)$, then $\Sigma |c_n|^2$ is convergent,

$$\frac{1}{2\pi}\int_{-\pi}^{\pi} |f|^2\,dx = \sum_{-\infty}^{\infty} |c_n|^2, \tag{2.7.1}$$

and

$$\frac{1}{2\pi}\int_{-\pi}^{\pi} |f-f_n|^2\,dx \to 0. \tag{2.7.2}$$

If F also is L^2 and $F \sim (C_n)$, then

$$\frac{1}{2\pi}\int_{-\pi}^{\pi} f\bar{F}\,dx = \sum_{-\infty}^{\infty} c_n \bar{C}_n. \tag{2.7.3}$$

If c_n is any sequence of numbers for which $\Sigma |c_n|^2$ is convergent, then there is an f of L^2, effectively unique, whose Fourier coefficients are the c_n.

Theorem 21. *There are corresponding results for the real trigonometrical system (T), the formulae corresponding to* (2.7.1) *and* (2.7.3), *when f and F are real, being*

$$\frac{1}{\pi}\int_{-\pi}^{\pi} f^2\,dx = \tfrac{1}{2}a_0^2 + \sum_{1}^{\infty}(a_n^2+b_n^2) \tag{2.7.4}$$

and

$$\frac{1}{\pi}\int_{-\pi}^{\pi} fF\,dx = \tfrac{1}{2}a_0A_0 + \sum_{1}^{\infty}(a_nA_n+b_nB_n). \tag{2.7.5}$$

2.8. Some theorems concerning other systems. In this and the next section we collect a number of theorems which can be deduced from Theorem 19. All these theorems are of the same depth, and their logical interdependence can be exhibited in many different ways.

(1) Our first remark is trivial. The systems (C) and (S) of § 1.10 (3) are plainly not complete in $(-\pi, \pi)$: thus all the F.c. of $\sin x$ with respect to (C) are 0. But each system is complete in $(0, \pi)$. Suppose, for example, that f is given in $(0, \pi)$ and that all its F.c. with respect

to (C) are 0; and let $f^*(x) = f(x)$ in $(0, \pi)$, $f^*(x) = f(-x)$ in $(-\pi, 0)$, so that f^* is even. Then all the F.c. of f^* with respect to (T) are 0, so that $f^* \equiv 0$ and therefore $f \equiv 0$.

(2) Our next theorem concerns a non-orthogonal system.

Theorem 22. *The system (x^n), for $n = N, N+1, \ldots$, is complete in $L(a, b)$.*

Here (a, b) is any finite interval. We have to prove that

$$\int_a^b f(x)\, x^n\, dx = 0 \quad (n = N, N+1, \ldots)$$

implies $f \equiv 0$. By considering $x^N f$ instead of f we can reduce the theorem to the case $N = 0$; and then, plainly, by a linear transformation to the case $a = -\pi$, $b = \pi$. But in this case, since the Taylor's series $\Sigma u_n x^n$ for $\cos mx$ and $\sin mx$ are uniformly convergent in $(-\pi, \pi)$, we have

$$\int_{-\pi}^{\pi} f(x) \begin{matrix}\cos\\ \sin\end{matrix} mx\, dx = \Sigma u_n \int_{-\pi}^{\pi} f(x)\, x^n\, dx = 0$$

for every m. Hence $f \equiv 0$, by Theorem 19.

A corollary is the completeness in $L(-1, 1)$ of the Legendre system of § 1.10 (4).

2.9. Weierstrass's theorem. The system (x^n) is complete in L, and *a fortiori* complete in any L^p and in C. It follows from Theorem 18 that it is closed in any L^p with $1 \leqq p < \infty$. But this inference depends upon a theorem which we have not proved, and in any case does not reveal the full truth, which is contained in a famous theorem of Weierstrass.

Theorem 23. *If $f(x)$ is continuous in $\langle a, b\rangle$, then there is a polynomial $P(x)$ such that $|f - P| < \epsilon$ for $a \leqq x \leqq b$.*

In other words, (x^n) is closed in C. There are a number of direct proofs of Weierstrass's theorem, all based on important principles; but it is most natural here to deduce it from the corresponding theorem for trigonometrical polynomials.

Theorem 24. *If $f(x)$ is periodic, and continuous in $\langle -\pi, \pi\rangle$, then there is a trigonometrical polynomial $T_n(x)$ such that $|f - T_n| < \epsilon$ for $-\pi \leqq x \leqq \pi$.*

There is plainly a periodic g of C_2 such that $|f-g| < \frac{1}{2}\epsilon$. For such a g, and for $n > 0$,

$$\int_{-\pi}^{\pi} g \frac{\cos}{\sin} nx\,dx = -\frac{1}{n^2}\int_{-\pi}^{\pi} g'' \frac{\cos}{\sin} nx\,dx = O\left(\frac{1}{n^2}\right),$$

by two partial integrations; so that the F.s. of g is uniformly convergent and, by Theorems 8 and 19, its sum is g. Hence $|g-T_n| < \frac{1}{2}\epsilon$ if T_n is the sum of a sufficient number of terms of the F.s. of g, and $|f-T_n| < \epsilon$.

Incidentally Theorem 24 shows (without use of Theorem 18) that (T) is closed in L^p. For the class C^* of continuous periodic functions is dense in L^p, by Theorem 1, and (T), by Theorem 24, is closed in C^*.

Theorem 23 is a corollary. We begin by reducing the theorem, by a linear substitution, to the case $a = -\frac{1}{2}\pi$, $b = \frac{1}{2}\pi$; and define f, outside $(-\frac{1}{2}\pi, \frac{1}{2}\pi)$, so as to become a continuous function with period 2π. There is then a trigonometrical polynomial T_n such that $|f-T_n| < \frac{1}{2}\epsilon$ for all x. But T_n can be expanded, by Taylor's theorem, in a power series converging uniformly in $\langle -\pi, \pi \rangle$. If P is the sum of a sufficient number of terms of this series, then $|T_n - P| < \frac{1}{2}\epsilon$, $|f-P| < \epsilon$, for $-\frac{1}{2}\pi \leqq x \leqq \frac{1}{2}\pi$.

III. FURTHER PROPERTIES OF TRIGONOMETRICAL FOURIER SERIES

3.1. Simple properties of Fourier constants. From this point on we concentrate our attention on the trigonometrical F.s. defined in § 1.3. We use the complex system (E) or the real system (T) as may be convenient: in the latter case we shall usually suppose f real. In any case we suppose f periodic. We begin with a few trivialities.

Theorem 25. *If $f(\theta) \sim (c_n)$, then $f(-\theta) \sim (c_{-n})$ and $\bar{f}(\theta) \sim (\bar{c}_{-n})$. If m is integral, then $e^{mi\theta} f(\theta) \sim (c_{n-m})$. If $f \sim (c_n)$, $g \sim (d_n)$, then*

$$af + bg \sim (ac_n + bd_n).$$

Theorem 26. *If $f(\theta) \sim (c_n)$, then $f(\theta+\alpha) \sim (C_n(\alpha))$, and if $f(\theta) \sim (a_n, b_n)$, then $f(\theta+\alpha) \sim (A_n(\alpha), B_n(\alpha))$.**

* A_n, B_n, and C_n were defined in (1.1.2), (1.2.6), and (1.1.8).

Theorem 27. *If f is absolutely continuous, then $f' \sim (inc_n)$ or $f' \sim (nb_n, -na_n)$.*

We leave the proofs to the reader: the last theorem is proved by partial integration, as in § 2.6.

Theorem 28. *If f and g are L^2, $f \sim (c_n)$, and $g \sim (d_n)$, then $fg \sim (r_n)$, where*

$$r_n = \sum_{-\infty}^{\infty} c_m d_{n-m}. \tag{3.1.1}$$

It follows from Theorem 25 that $\bar{g} \sim (\bar{d}_{-n})$ and $h = \bar{g}e^{mi\theta} \sim (\bar{d}_{m-n})$. Hence, applying Parseval's theorem, in the form (2.3.1), to f and h, we obtain

$$r_m = \frac{1}{2\pi}\int fge^{-mi\theta}\,d\theta = \frac{1}{2\pi}\int f\bar{h}\,d\theta = \Sigma c_n d_{m-n},$$

which is (3.1.1) with m and n interchanged. The expressions for the real F.c. of fg are a little more complicated.

The formula (3.1.1) resembles (1.8.1), and the sequence (r_n) may be called the resultant of (c_n) and (d_n). This correspondence suggests a reciprocal theorem.

Theorem 29. *If f and g are L, $f \sim (c_n)$, $g \sim (d_n)$, and r is the resultant of f and g, then $r \sim (c_n d_n)$.*

By Theorem 4, r is L. Also

$$\frac{1}{2\pi}\int r(\theta)\,e^{-ni\theta}\,d\theta = \frac{1}{2\pi}\int d\theta\left\{\frac{1}{2\pi}\int f(\theta-t)\,e^{-ni(\theta-t)}\,.\,g(t)\,e^{-nit}\,dt\right\}.$$

Here the inner integral is the resultant of $f(\theta)\,e^{-ni\theta}$ and $g(\theta)\,e^{-ni\theta}$; and therefore, by (1.8.2),

$$\frac{1}{2\pi}\int r(\theta)\,e^{-ni\theta}\,d\theta = \frac{1}{2\pi}\int f(\theta)\,e^{-ni\theta}\,d\theta\,\frac{1}{2\pi}\int g(\theta)\,e^{-ni\theta}\,d\theta = c_n d_n.$$

3.2. The Riemann-Lebesgue theorem. We now come to one of the fundamental theorems of the subject.

Theorem 30. *The Fourier constants of any integrable function tend to zero.*

This theorem is a special case of Theorem 15. Its importance justifies an alternative proof, and we arrange this proof so as to obtain a considerable generalization.

Theorem 31. *Suppose that f and g have period 2π; that f is L and g is V; that λ and α are real; and that $-\pi \leqq a \leqq b \leqq \pi$. Then*

$$J(a,b,\alpha,\lambda) = \int_a^b f(\theta+\alpha)\, g(\theta)\, e^{-\lambda i\theta}\, d\theta \to 0 \tag{3.2.1}$$

uniformly in a, b, and α, when $|\lambda| \to \infty$.

Theorem 30 is the case $a = -\pi$, $b = \pi$, $\alpha = 0$, $g = 1$, $\lambda = n$.

(i) Suppose first that $g = 1$. We may also suppose $\lambda > 0$. Then

$$J = \int_a^b f(\theta+\alpha)\, e^{-\lambda i\theta}\, d\theta = -\int_{a-\pi/\lambda}^{b-\pi/\lambda} f\left(\theta+\alpha+\frac{\pi}{\lambda}\right) e^{-\lambda i\theta}\, d\theta$$

$$= -\int_a^b f\left(\theta+\alpha+\frac{\pi}{\lambda}\right) e^{-\lambda i\theta}\, d\theta + o(1),$$

since the error introduced by changing the limits, which is at most $\int |f|\, d\theta$ over two intervals of length π/λ, is $o(1)$ uniformly. Hence

$$J = \tfrac{1}{2}\int_a^b \left\{f(\theta+\alpha) - f\left(\theta+\alpha+\frac{\pi}{\lambda}\right)\right\} e^{-\lambda i\theta}\, d\theta + o(1), \tag{3.2.2}$$

$$|J| \leqq \tfrac{1}{2}\int_a^b \left|f\left(\theta+\alpha+\frac{\pi}{\lambda}\right) - f(\theta+\alpha)\right| d\theta + o(1) = o(1), \tag{3.2.3}$$

by Theorem 3 (with $p = 1$).

If in particular $a = -\pi$, $b = \pi$, $\alpha = 0$, $\lambda = n$, then all the integrations may be taken over $(-\pi, \pi)$, and the $o(1)$ in (3.2.2) may be deleted. We thus obtain an important upper bound for $|c_n|$, viz.

$$|c_n| \leqq \frac{1}{4\pi}\int_{-\pi}^{\pi} \left|f\left(\theta+\frac{\pi}{n}\right) - f(\theta)\right| d\theta \quad (n \neq 0). \tag{3.2.4}$$

(ii) We may suppose f and g real. If g is V, then $g = g_1 - g_2$, where g_1 and g_2 are positive, bounded, and decreasing. Then J is composed of four integrals of similar type, one of which is e.g.

$$\int_a^b f(\theta+\alpha)\, g_1(\theta) \cos \lambda\theta\, d\theta = g_1(a)\int_a^c f(\theta+\alpha) \cos\lambda\theta\, d\theta,$$

where $a < c < b$: and the result follows from (i).

3.3. Some simple inequalities. In this section we collect some simple theorems concerning the F.c. of functions of special types.

Theorem 32. *If $f \geqq 0$, then $|c_n| \leqq c_0$, $|a_n| \leqq a_0$, $|b_n| \leqq a_0$.*

Theorem 33. *If f is odd, and $f \geqq 0$ in $(0, \pi)$, then* $|b_n| \leqq nb_1$.

We leave the proofs of these theorems to the reader: for Theorem 33, observe that $|\sin n\theta| \leqq n |\sin\theta|$.

Theorems 32 and 33 have interesting applications to the theory of analytic functions.

(i) Suppose that $F(z) = C_0 + C_1 z + C_2 z^2 + \ldots$ is an analytic function regular for $r < 1$, that $F = u - iv$, and that $u \geqq 0$. Then u, for a fixed r, satisfies the conditions of Theorem 32, and so, after (1.2.10), $|C_n| r^n \leqq 2\mathbf{R}C_0$. Making $r \to 1$, we see that *if $F = \Sigma C_n z^n$ has its real part non-negative for $r < 1$, then $\mathbf{R}C_0 \geqq 0$ and $|C_n| \leqq 2\mathbf{R}C_0$ for $n > 0$.*

(ii) Suppose that $F(z) = z + C_2 z^2 + \ldots$ is regular and schlicht* in $|z| < 1$, and that its coefficients are real. A little consideration shows that the image of the circle $|z| = r < 1$ by the transformation $w = F(z)$ is a Jordan curve with just two real points, viz. those corresponding to $z = \pm r$, and that v, for a fixed r, satisfies the conditions of Theorem 33. Hence, after (1.2.10), $|C_n| r^n \leqq nr$. Making $r \to 1$, we see that *if $F = z + C_2 z^2 + \ldots$ has real coefficients and is schlicht in $|z| < 1$, then $|C_n| \leqq n$.*

In the next two theorems it is better to take $(0, 2\pi)$ as the fundamental interval.

Theorem 34. *If f decreases throughout $(0, 2\pi)$, then* $b_n \geqq 0$.

For $$\pi b_n = \sum_{k=0}^{n-1} \int_0^{\pi/n} \left\{ f\left(\frac{2k\pi}{n} + \theta\right) - f\left(\frac{2k\pi}{n} + \frac{\pi}{n} + \theta\right) \right\} \sin n\theta \, d\theta \geqq 0.$$

Theorem 35. *If f is convex in $(0, 2\pi)$, and $n > 0$, then* $a_n \geqq 0$.

For
$$\pi a_n = \sum_0^{n-1} \int_0^{\pi/2n} \left\{ f\left(\frac{2k\pi}{n} + \theta\right) - f\left(\frac{2k\pi}{n} + \frac{\pi}{n} + \theta\right) - f\left(\frac{2k\pi}{n} + \frac{\pi}{n} - \theta\right) + f\left(\frac{2k\pi}{n} + \frac{2\pi}{n} - \theta\right) \right\} \cos n\theta \, d\theta \geqq 0,$$
since $f(x+h) - f(x)$ increases with x for every positive h.

3.4. The order of magnitude of Fourier constants. The Riemann-Lebesgue theorem shows that the F.c. of any f tend to zero; and this is in a sense the most that is true, even for continuous functions. For, if (χ_n) is any sequence of positive decreasing numbers whose limit is 0, we can choose a sequence of integers n_k tending to infinity so rapidly that $k^{-2} > \chi_{n_k}$. The series $\Sigma k^{-2} \cos n_k \theta$ is uniformly convergent, and so the F.s. of a continuous function for which $a_n = k^{-2} > \chi_n$ when $n = n_k$.

* A function is schlicht in a region if it does not assume the same value twice.

Theorem 36. *If* $|f(\theta+h)-f(\theta)|=O(|h|^\alpha)$, *where* $0<\alpha<1$, *uniformly in* θ, *or, more generally, if*

$$\mathbf{M}_p\{f(\theta+h)-f(\theta)\} = \left\{\frac{1}{2\pi}\int_{-\pi}^{\pi} |f(\theta+h)-f(\theta)|^p \, d\theta\right\}^{1/p} = O(|h|^\alpha),$$

where $p \geqq 1$, *then* $c_n = O(|n|^{-\alpha})$.

For $$|c_n| \leqq \tfrac{1}{2}\mathbf{M}_1\left\{f\left(\theta+\frac{\pi}{n}\right)-f(\theta)\right\} \leqq \tfrac{1}{2}\mathbf{M}_p\left\{f\left(\theta+\frac{\pi}{n}\right)-f(\theta)\right\},$$

by (3.2.4) and (1.5.5.). There is a corresponding theorem with o.

Theorem 37. *If* f *is* V, *then* $c_n = O(|n|^{-1})$. *More precisely,* $2\pi|nc_n| \leqq \mathrm{V}$, *where* V *is the total variation of* f *in* $(0, 2\pi)$.

Suppose, for example, that $n > 0$. Then

$$|c_n| = \left|\frac{1}{2\pi}\int_{-\pi}^{\pi} f(\theta)\, e^{-ni\theta}\, d\theta\right| = \left|\frac{1}{2n\pi i}\int_{-\pi}^{\pi} e^{-ni\theta}\, df(\theta)\right| \leqq \frac{\mathrm{V}}{2n\pi}.$$

Theorem 38. *If* f *is absolutely continuous (and in particular if* f' *is bounded), then* $c_n = o(|n|^{-1})$.

For, if $n \neq 0$, and c'_n is the nth F.c. of f', then $inc_n = c'_n = o(1)$, by Theorems 27 and 30.

Theorem 39. *If* f *is continuous in* $\langle -\pi, \pi\rangle$, *except at a finite number* J *of points* ξ_j *where there is a jump* $d_j = f(\xi_j+0)-f(\xi_j-0)$, *and is absolutely continuous in the intervals of continuity; and* $n \neq 0$; *then*

$$c_n + \frac{i}{2n\pi}\sum_{1}^{J} d_j e^{-ni\xi_j} = -\frac{i}{2n\pi}\int_{-\pi}^{\pi} f'(\theta)\, e^{-ni\theta}\, d\theta = o\left(\frac{1}{|n|}\right). \tag{3.4.1}$$

The proof, by sub-division of the interval and partial integration, is immediate.

Theorem 40. *If* $f^{(p-1)}$ *is absolutely continuous (and in particular if* $f^{(p)}$ *is bounded), then* $c_n = o(|n|^{-p})$.

This is another corollary of Theorems 27 and 30.

3.5. Functions of bounded variation. The most familiar functions of V are either discontinuous or absolutely continuous, and in the latter case their F.c. are $o(|n|^{-1})$. This might suggest that the F.c. of a function f of V are $o(|n|^{-1})$ if and only if f is continuous. Actually, the truth is less simple.

Theorem 41. *There are continuous functions of V for which* $c_n \neq o(|n|^{-1})$.

We take $(0, 2\pi)$ as the fundamental interval. We denote by $I_{1,1}$; $I_{2,1}$, $I_{2,2}$; $I_{3,1}$, $I_{3,2}$, ... the excluded intervals of 'Cantor's ternary set' in $(0, 2\pi)$: $I_{1,1}$ is $(\frac{2}{3}\pi, \frac{4}{3}\pi)$, the central third; $I_{2,1}$, $I_{2,2}$ the central thirds of the remaining intervals $(0, \frac{2}{3}\pi)$ and $(\frac{4}{3}\pi, 2\pi)$; and so on. We define $\phi(\theta)$ as $\frac{1}{2}$ in $I_{1,1}$; $\frac{1}{4}$, $\frac{3}{4}$ in $I_{2,1}$, $I_{2,2}$; $\frac{1}{8}$, $\frac{3}{8}$, ... in $I_{3,1}$, $I_{3,2}$, ...; and so on: this defines $\phi(\theta)$ for all θ except the θ of Cantor's set, in which it is defined by continuity. Then $\phi(\theta)$ is continuous, increases with θ from 0 to 1, and has a derivative 0 p.p. It is the standard example of a continuous, but not absolutely continuous, function of V. If $f(\theta) = \phi(\theta) - \theta/2\pi$ for $0 \leqq \theta \leqq 2\pi$, and is periodic, then $f(\theta)$ is a periodic and continuous function of V. If $n > 0$, then

(3.5.1)
$$c_n = \frac{1}{2\pi}\int_0^{2\pi}\left\{\phi(\theta) - \frac{\theta}{2\pi}\right\} e^{-ni\theta}\,d\theta = \frac{1}{2n\pi i}\int_0^{2\pi} e^{-ni\theta}\,d\phi(\theta) = \frac{p_n}{2n\pi i},$$

say. If we can show that p_n does not tend to 0, then $f(\theta)$ will satisfy the requirements of the theorem.

We take $n = 3^m$. Then

$$p_{3^m} = \int_0^{2\pi} e^{-3^m i\theta}\,d\phi(\theta) = 2\int_0^{\frac{2}{3}\pi} e^{-3^m i\theta}\,d\phi(\theta), \tag{3.5.2}$$

since ϕ is constant in $(\frac{2}{3}\pi, \frac{4}{3}\pi)$ and varies similarly in the two remaining intervals; and this is

$$2\int_0^{2\pi} e^{-3^{m-1}i\theta}\,d\phi(\tfrac{1}{3}\theta) = \int_0^{2\pi} e^{-3^{m-1}i\theta}\,d\phi(\theta) = p_{3^{m-1}},$$

because $\phi(\frac{1}{3}\theta) = \frac{1}{2}\phi(\theta)$. It follows that $p_{3^m} = p_1$, and it is therefore sufficient to prove that $p_1 \neq 0$. But $\phi(\theta)$ is constant in $(\frac{2}{9}\pi, \frac{4}{9}\pi)$, and therefore, after (3.5.2),

$$\mathbf{R}p_1 = 2\left(\int_0^{\frac{2}{9}\pi} + \int_{\frac{4}{9}\pi}^{\frac{2}{3}\pi}\right)\cos\theta\,d\phi(\theta) = 2\int_0^{\frac{2}{9}\pi}\{\cos\theta + \cos(\theta + \tfrac{4}{9}\pi)\}\,d\phi(\theta);$$

and $\cos\theta + \cos(\theta + \frac{4}{9}\pi) = 2\cos(\theta + \frac{2}{9}\pi)\cos\frac{2}{9}\pi \geqq 2\cos\frac{2}{9}\pi\cos\frac{4}{9}\pi$.

Hence $$\mathbf{R}p_1 \geqq 4\cos\tfrac{2}{9}\pi\cos\tfrac{4}{9}\pi\int_0^{\frac{2}{9}\pi} d\phi(\theta) = \cos\tfrac{2}{9}\pi\cos\tfrac{4}{9}\pi > 0.$$

A function of V may have an enumerable set of points of jump, and the values of f at these points do not affect its F.c. We may therefore suppose that $f(\theta) = \frac{1}{2}\{f(\theta+0)+f(\theta-0)\}$ for all θ; and then Wiener has shown that there is a simple necessary and sufficient condition for the continuity of f, viz.

$$\sum_{-n}^{n} |\, mc_m \,| = o(n).$$

This is true in particular if $c_n = o(|\, n\,|^{-1})$.

3.6. Some elementary formulae. It will be convenient to collect here some familiar theorems which we shall use repeatedly later. We leave the proofs to the reader.

First, there are the identities

(3.6.1)
$$D_n(\theta) = \tfrac{1}{2} + \cos\theta + \cos 2\theta + \ldots + \cos n\theta = \frac{\sin(n+\frac{1}{2})\theta}{2\sin\frac{1}{2}\theta},$$

(3.6.2)
$$\tilde{D}_n(\theta) = \sin\theta + \sin 2\theta + \ldots + \sin n\theta = \frac{\cos\frac{1}{2}\theta - \cos(n+\frac{1}{2})\theta}{2\sin\frac{1}{2}\theta},$$

(3.6.3)
$$F_n(\theta) = \frac{D_0(\theta)+D_1(\theta)+\ldots+D_n(\theta)}{n+1} = \frac{1-\cos(n+1)\theta}{4(n+1)\sin^2\frac{1}{2}\theta}.$$

These are true for all θ if we define the right-hand sides, when $\theta = 2k\pi$, so as to be continuous. Also

$$\text{(3.6.4)} \qquad \frac{1}{\pi}\int_{-\pi}^{\pi} D_n(\theta)\,d\theta = 1, \quad \frac{1}{\pi}\int_{-\pi}^{\pi} F_n(\theta)\,d\theta = 1.$$

Next, there are the inequalities

$$\text{(3.6.5)} \qquad |\sin\theta| \leqq |\theta|, \quad 1-\cos\theta \leqq \tfrac{1}{2}\theta^2,$$

$$\text{(3.6.6)} \qquad \sin\theta \geqq \frac{2}{\pi}\theta \quad (0 \leqq \theta \leqq \tfrac{1}{2}\pi), \quad 1-\cos\theta \geqq \frac{2}{\pi^2}\theta^2 \quad (0 \leqq \theta \leqq \pi),$$

$$\text{(3.6.7)} \qquad \left|\sum_{p}^{q} e^{ni\theta}\right| \leqq \frac{1}{|\sin\frac{1}{2}\theta|}, \quad \left|\sum_{p}^{q} \lambda_n e^{ni\theta}\right| \leqq \frac{\lambda_p}{|\sin\frac{1}{2}\theta|}.$$

In (3.6.7), $\theta \neq 2k\pi$, $0 \leqq p \leqq q$, and λ_n is positive and decreases as n increases. A corollary is that, if λ_n decreases to 0, then $\Sigma\lambda_n e^{ni\theta}$ is uniformly convergent in any closed interval free from multiples of 2π.

Finally, we define Δu_n and $\Delta^2 u_n$ by

$$\Delta u_n = u_n - u_{n+1}, \quad \Delta^2 u_n = \Delta(\Delta u_n).$$

If $u_n \to 0$, then $\Sigma \Delta u_n = u_0$; if also $\Delta^2 u_n \geqq 0$, i.e. if (u_n) is *convex*, then $\Delta u_n \geqq 0$, $n\Delta u_n \to 0$ and

$$\Sigma(n+1)\,\Delta^2 u_n = \Sigma \Delta u_n = u_0. \tag{3.6.8}$$

Here all the summations are from 0 to ∞.

3.7. A special trigonometrical series. There is one special series, viz.

$$\mathbf{S}(\theta) = \sum_1^\infty \frac{\sin n\theta}{n}, \tag{3.7.1}$$

which is particularly important in the general theory.

We define $\mathbf{f}(\theta)$ by

$$\mathbf{f}(\theta) = \tfrac{1}{2}(\pi - \theta) \quad (0 < \theta < 2\pi), \qquad \mathbf{f}(0) = 0, \tag{3.7.2}$$

and elsewhere by periodicity. It is odd and has a jump π at 0, and is the simplest periodic function with just one jump in a period.

If $\mathbf{f}(\theta) \sim (a_n, b_n)$, then $a_n = 0$ and

$$b_n = \frac{1}{\pi}\int_0^\pi (\pi - \theta)\sin n\theta\, d\theta = \frac{1}{n},$$

by partial integration, if $n > 0$. Thus the F.s. of $\mathbf{f}$ is $\mathbf{S}$.

The series $\mathbf{S}$ is convergent for all θ, and uniformly convergent in any closed interval free from multiples of 2π. We shall now prove that it is *boundedly* convergent. More generally, we prove that *if λ_n is a positive decreasing function of n, and $n\lambda_n < H$, then $\Sigma \lambda_n \sin n\theta$ is boundedly convergent*. We may suppose that $H = 1$ and $0 < \theta < \pi$. Then, if $\nu = \text{Min}\,(n, [\pi/\theta])$, we have

$$s_n(\theta) = \sum_1^n \lambda_m \sin m\theta = \left(\sum_1^\nu + \sum_{\nu+1}^n\right)\lambda_m \sin m\theta = s_n^{(1)}(\theta) + s_n^{(2)}(\theta)$$

(the second sum being empty if $\nu \geqq n$). Here

$$|\, s_n^{(1)}(\theta)\,| \leqq \theta \sum_1^\nu m\lambda_m \leqq \nu\theta \leqq \pi,$$

and

$$|\, s_n^{(2)}(\theta)| \leqq \frac{\lambda_{\nu+1}}{\sin\frac{1}{2}\theta} \leqq \frac{\pi}{(\nu+1)\,\theta} \leqq 1,$$

by (3.6.5)–(3.6.7).

In particular, $\mathbf{S}$ is boundedly convergent; and therefore, by Theorem 8, its sum is $\mathbf{f}$. We have thus proved

Theorem 42. *The series* **S** *is the Fourier series of* **f**, *and converges to* **f**, *boundedly, and uniformly in any closed interval free from multiples of* 2π.

Our proof of Theorem 42 depends (owing to our appeal to Theorem 8) on the completeness of the trigonometrical system (T). It is therefore important to observe that **S** can be summed directly in other ways, which will be found in text-books of elementary analysis. For example, starting from the series $\log(1-z) = -z-\frac{1}{2}z^2-\ldots$, where $z = re^{i\theta}$, $r<1$, and $0<\theta<2\pi$, taking real and imaginary parts, and using Abel's theorem, we find that

$$(3.7.3)\quad \sum_1^\infty \frac{\sin n\theta}{n} = \tfrac{1}{2}(\pi-\theta), \quad \sum_1^\infty \frac{\cos n\theta}{n} = \log\frac{1}{2\sin\frac{1}{2}\theta} \quad (0<\theta<2\pi).$$

The first equation is part of Theorem 42. The second may be proved much as we proved the first, but the cosine series diverges when $\theta = 2k\pi$, and we have to prove it *dominatedly* convergent in $(0, 2\pi)$.

A simple corollary is that

$$(3.7.4)\qquad \sin\theta+\tfrac{1}{3}\sin 3\theta+\tfrac{1}{5}\sin 5\theta+\ldots = \tfrac{1}{4}\pi\operatorname{sgn}\theta \quad (-\pi<\theta<\pi),$$

where $\operatorname{sgn}\theta$ is 1, 0, or -1 according as θ is positive, zero, or negative. The series is boundedly convergent, and is the F.s. of its sum, which has jumps at all points $k\pi$, and is the simplest periodic step-function which is not constant.

3.8. Integration of Fourier series.

We now apply Theorem 42 to the proof of an important general theorem.

Theorem 43. *If* $f \sim (a_n, b_n)$, *then*

$$(3.8.1)\quad \sum_1^\infty \frac{b_n}{n} = \frac{1}{2\pi}\int_0^{2\pi} f(\theta)(\pi-\theta)\,d\theta = \frac{1}{2\pi}\int_{-\pi}^{\pi} f(\theta)(\pi\operatorname{sgn}\theta-\theta)\,d\theta.$$

Also

$$(3.8.2)\quad F(\theta) = \int_0^\theta f(t)\,dt-\tfrac{1}{2}a_0\theta = \sum_1^\infty \frac{b_n}{n}-\sum_1^\infty \frac{b_n\cos n\theta-a_n\sin n\theta}{n},$$

the last series being uniformly convergent.

It is plain that $F(\theta)$ is periodic and $F(2k\pi) = 0$. Also

$$(3.8.3)\quad \sum_1^n \frac{b_m}{m} = \frac{1}{\pi}\int_0^{2\pi} f(\theta)\sum_1^n \frac{\sin m\theta}{m}\,d\theta \to \frac{1}{2\pi}\int_0^{2\pi} f(\theta)(\pi-\theta)\,d\theta,$$

by Theorem 42, and this is (3.8.1).

Next, in proving (3.8.2), we may suppose that $a_0 = 0$. The F.c. of $f(\theta+t)$, as a function of t, are $A_n(\theta)$ and $B_n(\theta)$, by Theorem 26.

Hence, applying what we have proved already to $f(\theta+t) = F'(\theta+t)$, we find that

$$\sum_1^\infty \frac{B_n(\theta)}{n} = \frac{1}{2\pi}\int_0^{2\pi} F'(\theta+t)(\pi-t)\,dt = -F(\theta)+\frac{1}{2\pi}\int_0^{2\pi} F(\theta+t)\,dt.$$

Since

$$\frac{1}{2\pi}\int_0^{2\pi} F(\theta+t)\,dt = \frac{1}{2\pi}\int_0^{2\pi} F(t)\,dt = \frac{1}{2\pi}\int_0^{2\pi} f(t)(\pi-t)\,dt = \sum_1^\infty \frac{b_n}{n},$$

by partial integration and (3.8.1), this is equivalent to (3.8.2).

Finally, the series (3.8.2) is uniformly convergent. For

$$\left|\sum_p^q \frac{B_n(\theta)}{n}\right| \leqq \frac{1}{\pi}\int_0^{2\pi} |f(\theta+t)| \left|\sum_p^q \frac{\sin nt}{n}\right| dt$$

$$\leqq H\left(\int_0^\delta + \int_{2\pi-\delta}^{2\pi}\right)|f(\theta+t)|\,dt + \frac{1}{\pi}\int_\delta^{2\pi-\delta} |f(\theta+t)| \left|\sum_p^q \frac{\sin nt}{n}\right| dt = I_1 + I_2,$$

say. We can choose $\delta = \delta(\epsilon)$ so that $|I_1| < \epsilon$; and then, since **S** is uniformly convergent in $(\delta, 2\pi-\delta)$, choose $N = N(\delta,\epsilon) = N(\epsilon)$ so that $|I_2| < \epsilon$ for $q > p \geqq N$.

We can also write (3.8.2) as

$$F(\theta) = \sum_1^\infty \frac{a_n \sin n\theta + b_n(1-\cos n\theta)}{n} = \sum_1^\infty \int_0^\theta A_n(t)\,dt.$$

We have therefore proved incidentally

Theorem 44. *Any Fourier series may be integrated term by term, and the integrated series is uniformly convergent.*

In 'complex' notation (3.8.2) is

$$(3.8.4) \qquad F(\theta) = \int_0^\theta f(t)\,dt - c_0\theta = i\sum_{n\neq 0} \frac{c_n}{n} - i\sum_{n\neq 0} \frac{c_n}{n} e^{ni\theta}.$$

Theorems 43 and 44 give an alternative proof of the completeness of (T). For if $a_n = b_n = 0$ for all n, then $F(\theta) = 0$ and so $f(\theta) \equiv 0$. The proof naturally presupposes that we have found the sum of **S** by some elementary method, without appealing to Theorem 8.

If we write t for θ in the first series (3.7.3), and integrate from $t = 0$ to $t = \theta$, we obtain

$$(3.8.5) \qquad \sum_1^\infty \frac{1-\cos n\theta}{n^2} = \tfrac{1}{2}\pi\theta - \tfrac{1}{4}\theta^2 \quad (0 \leqq \theta \leqq 2\pi).$$

If we integrate the second series from 0 to π, we obtain

$$\int_0^{\pi} \log(2\sin\tfrac{1}{2}\theta)\,d\theta = 0,$$

and so

$$\int_0^{\frac{1}{2}\pi} \theta\cot\theta\,d\theta = -\int_0^{\frac{1}{2}\pi} \log\sin\theta\,d\theta = \tfrac{1}{2}\pi\log 2. \tag{3.8.6}$$

3.9. An elementary convergence theorem. We can use Theorems 39 and 43 to obtain a theorem about the convergence of F.s. which covers a large proportion of the series occurring in elementary work.

Theorem 45. *If $f(\theta)$ satisfies the conditions of Theorem 39, then its F.s. converges boundedly to $\frac{1}{2}\{f(\theta+0)+f(\theta-0)\}$, and uniformly, to $f(\theta)$, in any closed interval of continuity.*

It follows from (3.4.1) that

$$\sum c_n e^{ni\theta} = c_0 - \frac{i}{2\pi}\sum_1^J d_j \sum_{n\neq 0} \frac{1}{n} e^{ni(\theta-\xi_j)} - i\sum_{n\neq 0} \frac{\gamma_n}{n} e^{ni\theta}$$

(as a formal series), γ_n being the F.c. of f'. The last series converges uniformly, by Theorem 43, since f' is L. The second term on the right may be written as

$$\frac{1}{\pi}\sum_1^J d_j \sum_1^{\infty} \frac{\sin n(\theta-\xi_j)}{n} = \psi(\theta);$$

it converges boundedly, and uniformly in any closed interval of continuity of its sum; and its sum at ξ_j is

$$\psi(\xi_j) = \tfrac{1}{2}\{\psi(\xi_j+0)+\psi(\xi_j-0)\}.$$

The theorem now follows from Theorem 8.

3.10. Series with decreasing coefficients. The series

$$\text{(C)}\quad \tfrac{1}{2}\lambda_0 + \Sigma\lambda_n\cos n\theta, \qquad \text{(S)}\quad \Sigma\lambda_n\sin n\theta,$$

in which λ_n is positive and decreases steadily to 0, are uniformly convergent in any interval $0<\delta\leqq\theta\leqq 2\pi-\delta$, to sums $f(\theta)$ and $g(\theta)$: S is convergent for all θ, C except perhaps for $\theta = 2k\pi$. They have also a number of interesting and less obvious properties, of which we now give some account. Since C is even and S odd, we need only consider the interval $(0,\pi)$.

Theorem 46. *If f (or g) is L, then* C *(or* S*) is the Fourier series of f (or g).*

This is a corollary of a rather difficult general theorem (Theorem 100) which we shall prove in Ch. VII. There are however simple proofs for these special series.

(i) The series $\Sigma\lambda_n \sin n\theta \sin m\theta$ is uniformly convergent in $\langle 0, \pi\rangle$, for any fixed m, since

$$\left|\sum_{n=p}^{q} \lambda_n \sin n\theta \sin m\theta\right| \leqq m\theta \left|\sum_{n=p}^{q} \lambda_n \sin n\theta\right| \leqq \frac{m\theta\lambda_p}{\sin\frac{1}{2}\theta} \leqq \pi m\lambda_p,$$

by (3.6.5), (3.6.6), and (3.6.7); and its sum is $g(\theta)\sin m\theta$. Hence

$$\frac{2}{\pi}\int_0^{\pi} g(\theta)\sin m\theta\, d\theta = \lambda_m.$$

(ii) Similarly, using the second inequality (3.6.5), we see that

$$\tfrac{1}{2}\lambda_0(1-\cos m\theta)+\Sigma\lambda_n \cos n\theta(1-\cos m\theta)$$

converges uniformly to $f(\theta)\,(1-\cos m\theta)$, so that

$$\frac{2}{\pi}\int_0^{\pi} f(\theta)\, d\theta - \frac{2}{\pi}\int_0^{\pi} f(\theta)\cos m\theta\, d\theta = \lambda_0 - \lambda_m.$$

Making $m \to \infty$, and using Theorem 30, we see (first for $m = 0$ and then generally) that λ_m is the F.c. of f of rank m.

Theorem 47. *If*

$$\Lambda = \sum_1^{\infty} \frac{\lambda_n}{n} < \infty, \tag{3.10.1}$$

then C *and* S *are Fourier series. For* S, *the condition is also necessary.*

If $\Lambda_k = \frac{1}{2}\lambda_0 + \lambda_1 + \ldots + \lambda_k$, then

$$\sum_1^{\infty} \frac{\Lambda_k}{k(k+1)} = \tfrac{1}{2}\lambda_0 + \sum_1^{\infty} \frac{1}{k(k+1)} \sum_1^{k} \lambda_m = \tfrac{1}{2}\lambda_0 + \sum_1^{\infty} \lambda_m \sum_m^{\infty} \frac{1}{k(k+1)} = \tfrac{1}{2}\lambda_0 + \Lambda.$$

Hence, if $\pi/(k+1) \leqq \theta < \pi/k$ and $h = f+ig$, we have

$$h = \tfrac{1}{2}\lambda_0 + \sum_1^{\infty} \lambda_n e^{ni\theta} = \tfrac{1}{2}\lambda_0 + \sum_1^{k-1} \lambda_n e^{ni\theta} + \sum_k^{\infty} \lambda_n e^{ni\theta}.$$

$$|\,h\,| \leqq \Lambda_k + \frac{\lambda_k}{\sin\frac{1}{2}\theta} \leqq \Lambda_k + \frac{\pi\lambda_k}{\theta} \leqq \Lambda_k + (k+1)\,\lambda_k,$$

$$\int_0^{\pi} |\,h\,|\, d\theta = \sum_1^{\infty} \int_{\pi/(k+1)}^{\pi/k} |\,h\,|\, d\theta$$

$$\leqq \sum_1^{\infty} \frac{\pi}{k(k+1)}\{\Lambda_k + (k+1)\,\lambda_k\} = \pi(\tfrac{1}{2}\lambda_0 + 2\Lambda).$$

If g is L, then $\Lambda < \infty$, by Theorem 46 and (3.8.1). The next theorem shows that C may be a F.s. when $\Lambda = \infty$.

Theorem 48. *If λ_n is convex, then f is non-negative and integrable, and* C *is the Fourier series of f.*

In this case $\lambda_n \geqq 0$, $\Delta\lambda_n \geqq 0$, and $\Delta^2\lambda_n \geqq 0$. Summing twice partially, and using (3.6.1) and (3.6.3), we find that

$$
\begin{aligned}
C_n(\theta) &= \tfrac{1}{2}\lambda_0 + \lambda_1 \cos\theta + \ldots + \lambda_n \cos n\theta \\
&= \frac{1}{2\sin\frac{1}{2}\theta}\left\{\sum_{k=0}^{n-1} \Delta\lambda_k \sin(k+\tfrac{1}{2})\theta + \lambda_n \sin(n+\tfrac{1}{2})\theta\right\} \\
&= \frac{1}{4\sin^2\frac{1}{2}\theta}\sum_{k=0}^{n-2} \Delta^2\lambda_k\{1-\cos(k+1)\theta\} \\
&\qquad\qquad + \Delta\lambda_{n-1}\frac{1-\cos n\theta}{4\sin^2\frac{1}{2}\theta} + \lambda_n\frac{\sin(n+\frac{1}{2})\theta}{2\sin\frac{1}{2}\theta}
\end{aligned}
$$

for $0 < \theta < 2\pi$. The last two terms tend to 0 when $n \to \infty$, so that

$$
f(\theta) = \frac{1}{4\sin^2\frac{1}{2}\theta}\sum_{0}^{\infty} \Delta^2\lambda_n\{1-\cos(n+1)\theta\},
$$

a series of non-negative terms. It follows that $f(\theta) \geqq 0$, and that

$$
\begin{aligned}
\int_0^\pi f(\theta)\,d\theta &= \tfrac{1}{4}\sum_{0}^{\infty} \Delta^2\lambda_n \int_0^\pi \frac{1-\cos(n+1)\theta}{\sin^2\frac{1}{2}\theta}\,d\theta \\
&= \tfrac{1}{2}\pi \sum_{0}^{\infty} (n+1)\Delta^2\lambda_n = \tfrac{1}{2}\pi\lambda_0,
\end{aligned}
$$

by (3.6.4) and (3.6.8).

Thus

$$
\sum_{0}^{\infty} \frac{\cos n\theta}{\log(n+2)} \quad \left(\text{or } \sum_{2}^{\infty} \frac{\cos n\theta}{\log n}\right) \tag{3.10.2}
$$

is a F.s. Theorem 47 shows that the corresponding sine series is not, since $\Sigma(n\log n)^{-1}$ is divergent. These examples are particularly important because they show that *the series conjugate to a F.s. is not necessarily itself a F.s.*

3.11. Series with decreasing coefficients (*continued*). The theorems of § 3.10 bear, roughly, on the integrability of f or g. There is another interesting group of theorems concerning their boundedness or continuity.

If a t.s. is uniformly (boundedly) convergent, its sum is continuous (bounded). We shall prove in § 6.3 that the converse is true for series with *positive* coefficients, so that, for such series, uniform (bounded) convergence is equivalent to continuity (boundedness) of the sum: in particular this is true for our present series C and S. It follows that if f is bounded then $\Sigma\lambda_n$ is convergent, in which case C is uniformly convergent and f continuous. There is more to be said about S and g, and we prove it independently of § 6.3.

Theorem 49. *The series* S *is uniformly (boundedly) convergent whenever its sum is continuous (bounded). The condition*

$$(3.11.1) \qquad \lambda_n = o(n^{-1}) \quad [\lambda_n = O(n^{-1})]$$

is necessary and sufficient either for uniform (bounded) convergence or for continuity (boundedness).

We prove the part of the theorem which concerns continuity: the other part* is a simpler variant. It is plainly enough to prove (3.11.1) *sufficient* for uniform convergence and *necessary* for continuity.

(i) Suppose that $\lambda_n = o(n^{-1})$ and $k = [\pi/\theta]$. We choose N so that $|n\lambda_n| < \epsilon$ for $n \geqq N$, and take $q > p \geqq N$. If $p \leqq k < q$, then

$$U = \sum_p^q \lambda_n \sin n\theta = \sum_p^k + \sum_{k+1}^q = U_1 + U_2,$$

$$|U_1| \leqq \theta \sum_p^k n\lambda_n < k\theta\epsilon \leqq \pi\epsilon,$$

$$|U_2| \leqq \lambda_{k+1} \operatorname{cosec} \tfrac{1}{2}\theta \leqq \pi\lambda_{k+1}/\theta \leqq (k+1)\lambda_{k+1} < \epsilon,$$

and so $|U| < (\pi+1)\epsilon$. If $k \geqq q$ or $k < p$, then the same conclusion follows from the argument applied to U_1 or U_2. Hence $|U| < (\pi+1)\epsilon$ in any case, and S is uniformly convergent.

(ii) If $g \to 0$ when $\theta \to 0$, then

$$\Sigma \frac{\lambda_n}{n}(1 - \cos n\theta) = \int_0^\theta g(t)\,dt = o(\theta).$$

But when $\theta = \pi/k$ the sum is at least

$$\sum_{\frac{1}{2}k}^{k} \frac{\lambda_n}{n}(1 - \cos n\theta) \geqq \frac{2\theta^2}{\pi^2} \sum_{\frac{1}{2}k}^{k} n\lambda_n \geqq \frac{2\theta^2}{\pi^2} (\tfrac{1}{2}k)^2 \lambda_k = \tfrac{1}{2}\lambda_k,$$

by (3.6.6). Hence $\lambda_k = o(\theta) = o(k^{-1})$.

* Already proved partially in § 3.7.

3.12. The Gibbs phenomenon. If $s_n(\theta)$ is the partial sum of the F.s. of f, we define the *Gibbs set* of f at ξ as the aggregate of values of c for which $s_n(\theta) \to c$ when $n \to \infty$ and $\theta \to \xi$ through appropriate sequences. We may also define it as the aggregate of limits

$$\lim s_n(\xi + h_n) = c,$$

(h_n) being a sequence such that $h_n \to 0$ and $nh_n \to a$, where $-\infty \leqq a \leqq \infty$. It is clear from Theorem 30, and considerations of continuity, that the Gibbs set is a finite or infinite interval $\langle \eta_1, \eta_2 \rangle$.

Suppose in particular that f is continuous in an interval including ξ, except at ξ itself, where it has a jump $d = f(\xi+0) - f(\xi-0)$, that $f(\xi) = \frac{1}{2}\{f(\xi+0) + f(\xi-0)\}$, and that the F.s. of f converges to f in this interval. Then

$$-\infty \leqq \eta_1 \leqq \underline{f} < \bar{f} \leqq \eta_2 \leqq \infty,$$

where $\underline{f}$ and $\bar{f}$ are the lesser and greater of $f(\xi \pm 0)$. It would be natural to expect that η_1 and η_2 should be $\underline{f}$ and $\bar{f}$, but we shall see that this is not so even in the simplest cases.

We consider first the special series **S**, for which f is the **f** of (3.7.2) and $\xi = 0$, $f(\xi - 0) = -\frac{1}{2}\pi$, $f(\xi+0) = \frac{1}{2}\pi$, $d = \pi$. We may suppose h_n positive. Then

(3.12.1)

$$s_n(h_n) = \sum_1^n \frac{\sin mh_n}{m} = \int_0^{h_n} \left(\sum_1^n \cos m\theta \right) d\theta = \int_0^{h_n} \frac{\sin (n+\frac{1}{2})\theta}{2 \sin \frac{1}{2}\theta} d\theta - \tfrac{1}{2}h_n$$

$$= \int_0^{h_n} \frac{\sin(n+\frac{1}{2})\theta}{\theta} d\theta + \int_0^{h_n} \left(\frac{1}{2\sin\frac{1}{2}\theta} - \frac{1}{\theta} \right) \sin(n+\tfrac{1}{2})\theta \, d\theta - \tfrac{1}{2}h_n.$$

The last two terms tend to 0 (the second by Theorem 31), and

$$(3.12.2) \quad \int_0^{h_n} \frac{\sin(n+\frac{1}{2})\theta}{\theta} d\theta = \int_0^{(n+\frac{1}{2})h_n} \frac{\sin t}{t} dt \to \int_0^a \frac{\sin t}{t} dt$$

when $nh_n \to a$. The last integral is positive (except for $a = 0$); has maxima for $a = (2k+1)\pi$ and minima for $a = 2k\pi$; and is $\frac{1}{2}\pi$ for $a = \infty$. Its absolute maximum is

$$(3.12.3) \qquad G = \int_0^{\pi} \frac{\sin t}{t} dt = 1{\cdot}85\ldots > \tfrac{1}{2}\pi.$$

Thus the Gibbs set of **f** *is the interval of length* $2G > \pi$ *centred round the origin.*

Passing to the general case, we write

$$(3.12.4) \quad g(\theta) = f(\theta) - \frac{d}{\pi} \sum_1^{\infty} \frac{\sin n(\theta - \xi)}{n} = f(\theta) - \frac{d}{\pi} \mathbf{f}(\theta - \xi),$$

so that $g(\xi) = f(\xi) = \frac{1}{2}\{f(\xi+0)+f(\xi-0)\}$. Then

$$g(\xi \pm 0) = f(\xi \pm 0) \mp \tfrac{1}{2}d = \tfrac{1}{2}\{f(\xi+0)+f(\xi-0)\} = g(\xi),$$

so that g is continuous at ξ. Let us suppose that, as happens in the simplest cases, the F.s. of g is uniformly convergent in an interval round ξ. Then $s_n(\xi+h_n)$, for g, tends to $g(\xi)$, and the Gibbs set of g consists of this one point. It follows that, for f,

$$s_n(\xi+h_n) \to \tfrac{1}{2}\{f(\xi+0)+f(\xi-0)\}+\frac{f(\xi+0)-f(\xi-0)}{\pi}\int_0^a \frac{\sin t}{t}\,dt,$$

when $h_n \to 0$ and $nh_n \to a$, and that the Gibbs set of f is the interval of length $2dG/\pi$ centred round $\frac{1}{2}\{f(\xi+0)+f(\xi-0)\}$.

We return to this subject in § 6.7.

IV. CONVERGENCE OF FOURIER SERIES

4.1. Introduction. We have already proved one or two theorems concerning the convergence of the F.s. of functions of special types. In this chapter we discuss the problem of convergence more systematically.

One preliminary remark is advisable. The 'convergence problem' seems at first sight the central and most natural problem of the theory, and it was the first to be discussed seriously, but it has lost a good deal of its importance as the result of later research. A series may 'converge' in many senses, of which the classical sense of Cauchy is only one; and some of these senses, such as the 'strong convergence' of Ch. II and the 'summability' of Ch. V, can be correlated more naturally with the most obvious characteristics of the generating function.

We denote the partial sum of $T(f)$ by $s_n(\theta)$ or $s_n(\theta, f)$. We write

$$\phi(t) = \phi(\theta, t) = \tfrac{1}{2}\{f(\theta+t)+f(\theta-t)\}. \tag{4.1.1}$$

The 'natural' sum for $T(f)$ is $\phi(+0)$, when this limit exists (as it does, for example, at a point of continuity or jump). The F.c. of $\phi(t)$, considered as a function of t, are $A_n(\theta)$ and 0, by Theorem 26; and the F.s. of $f(t)$, for $t = \theta$, is the F.s. of $\phi(t)$, which is a cosine series in t, for $t = 0$. This remark enables us, if we wish, to reduce any convergence problem for a F.s., at the point θ, to the special case in which $\theta = 0$ and the series is a cosine series.

4.2. The convergence problem for the Fourier series. We have

(4.2.1)

$$s_n(\theta)=\frac{1}{\pi}\int_{-\pi}^{\pi} f(t)\left\{\tfrac{1}{2}+\sum_1^n \cos m(t-\theta)\right\}dt=\frac{1}{2\pi}\int_{-\pi}^{\pi} f(t)\frac{\sin(n+\frac{1}{2})(t-\theta)}{\sin\frac{1}{2}(t-\theta)}\,dt$$

$$=\frac{1}{\pi}\int_0^{\pi}\phi(\theta,t)\frac{\sin(n+\frac{1}{2})t}{\sin\frac{1}{2}t}\,dt=\frac{2}{\pi}\int_0^{\pi}\phi(\theta,t)\,D_n(t)\,dt,$$

$D_n(t)$, 'Dirichlet's kernel', being defined as in (3.6.1). In particular, if we take $f(t)=c$, then $s_n(\theta)=c$ for all n; and, subtracting the result from (4.2.1), we obtain

$$s_n(\theta)-c=\frac{2}{\pi}\int_0^{\pi} g_c(t)\,D_n(t)\,dt, \tag{4.2.2}$$

where

$$g_c(t)=\phi(\theta,t)-c=\tfrac{1}{2}\{f(\theta+t)+f(\theta-t)-2c\}. \tag{4.2.3}$$

Here c may depend on θ.

Theorem 50. *In order that the Fourier series of $f(t)$, for $t=\theta$, should converge to c, it is necessary and sufficient that*

$$J=J(\delta,c,\lambda)=\int_0^{\delta} g_c(t)\frac{\sin\lambda t}{t}\,dt\to 0, \tag{4.2.4}$$

where $g_c(t)$ is defined by (4.2.3), δ is fixed, $0<\delta<\pi$, and λ is a continuous parameter which tends to infinity. In order that the series should be uniformly convergent, or that its partial sums should be bounded, in a set E in which $c=c(\theta)$ is bounded, it is necessary and sufficient that J should tend uniformly to 0, or be bounded, in E.*

After (4.2.2), $s_n(\theta)-c$ is equal to

$$\frac{2}{\pi}\int_0^{\delta} g\frac{\sin(n+\frac{1}{2})t}{t}\,dt+\frac{2}{\pi}\int_0^{\delta} g\left(\frac{1}{2\sin\frac{1}{2}t}-\frac{1}{t}\right)\sin(n+\tfrac{1}{2})t\,dt$$
$$+\frac{2}{\pi}\int_{\delta}^{\pi} g\frac{\sin(n+\frac{1}{2})t}{2\sin\frac{1}{2}t}\,dt.$$

The last two integrals are $o(1)$, by Theorem 31, and uniformly in any set in which $c(\theta)$ is bounded. Thus $s_n\to c$ is equivalent to $J(\delta,c,n+\frac{1}{2})\to 0$, and it remains only to show that we can replace $n+\frac{1}{2}$ by λ. But if $n-\frac{1}{2}<\lambda<n+\frac{1}{2}$ and

$$\mu=\tfrac{1}{2}(\lambda-n+\tfrac{1}{2}),\quad \nu=\tfrac{1}{2}(\lambda+n-\tfrac{1}{2}),$$

* When δ occurs in a theorem of this character, it is to be thought of as small: we shall not repeat that $0<\delta<\pi$. We shall often write $g(t)$, or simply g, for $g_c(t)$.

then $(\sin \mu t)/t$ decreases in $(0, \delta)$, and so

$$\int_0^\delta g \frac{\sin \lambda t - \sin (n-\frac{1}{2})t}{t} dt = 2\int_0^\delta g \frac{\sin \mu t}{t} \cos \nu t \, dt = 2\mu \int_0^\tau g \cos \nu t \, dt,$$

for a $\tau = \tau(\lambda)$ between 0 and δ; and this is $o(1)$ uniformly in any set in which c is bounded.

It will usually happen, as here, that our main theorem concerning the convergence of $s_n(\theta)$ can be supplemented by a clause concerning uniform convergence or boundedness: for example, the convergence will be uniform when the conditions of the theorem are satisfied uniformly in an interval, and $c(\theta)$ is bounded in that interval. We shall not usually give full proofs of these supplementary clauses, and sometimes we shall not even state them explicitly.

Theorem 51. *A sufficient condition that $s_n(\theta) \to c$ is that there should be an $\eta = \eta(\epsilon)$ and a $\Lambda = \Lambda(\epsilon)$ such that $|J(\eta, c, \lambda)| < \epsilon$ for $\lambda \geqq \Lambda$.*

For then
$$|J(\delta, c, \lambda)| < \epsilon + \left| \int_\eta^\delta g \frac{\sin \lambda t}{t} dt \right|,$$
and
$$\overline{\lim} \, |J(\delta, c, \lambda)| \leqq \epsilon + \lim \left| \int_\eta^\delta g \frac{\sin \lambda t}{t} dt \right| = \epsilon,$$
by Theorem 31.

We end this section with what is usually called 'Riemann's localization theorem'.

Theorem 52. *The behaviour of $s_n(\theta)$ when $n \to \infty$ depends only on the behaviour of $f(t)$ near $t = \theta$.*

This is naturally included in Theorems 50 and 51, since δ is arbitrary and the values of f outside $(\theta - \delta, \theta + \delta)$ are irrelevant.

4.3. Continuity conditions at a point. We shall usually find that our criteria for convergence contain two clauses, one of which is a 'continuity condition' varying little in different criteria, the other a sharper condition of a more individual kind.

A function $h(t)$, defined only for $t > 0$, is 'continuous for $t = 0$' if $h(t)$ tends to a limit $h(+0)$: we may then define $h(0)$ as $h(+0)$ if we please. Continuity in this sense may be destroyed by changing $h(t)$ in a null set, a change which does not affect the F.s. of a function. It is therefore natural, in this theory, to consider more general modes of continuity. We are concerned here with two generalized senses in which $h(t)$ may 'tend to 0'.

If

$$H(t) = \int_0^t h(u)\,du = o(t) \tag{4.3.1}$$

when $t \to 0$, we may say that $h(t) \to 0$ *in mean*. If $|h(t)| \to 0$ in mean, i.e. if

$$H^*(t) = \int_0^t |h(u)|\,du = o(t), \tag{4.3.2}$$

we may say that $h(t) \to 0$ *strongly in mean*. It is plain that (4.3.2) implies (4.3.1), but that the converse is false.

In our applications to F.s., $h(t)$ will be the function $g_c(t)$ defined in (4.2.3). If then $h(t)$ satisfies (4.3.1) or (4.3.2), for a certain $c(\theta)$, we shall say that $f(t)$ *satisfies* l_c *or* L_c *for* $t = \theta$ (the capital letter referring to the stronger condition with $|h|$).

It is plain that $f(t)$ satisfies L_c, and *a fortiori* l_c, with $c = \phi(+0)$, at a point of continuity or jump: but much more than this is true. It was proved by Lebesgue that, if f is L, and $\alpha(\theta)$ is any function of θ finite for almost all θ, then

$$\frac{1}{t}\int_0^t |f(\theta+u) - \alpha(\theta)|\,du \to |f(\theta) - \alpha(\theta)|$$

for almost all θ. It follows in particular that *f satisfies L_c (and a fortiori l_c), with $c = f(\theta)$, p.p.* We shall write L_f and l_f for these special forms of L_c and l_c.

If h satisfies (4.3.1), then

$$\int_0^{\pi/\lambda} h(t)\frac{\sin \lambda t}{t}\,dt = o(1). \tag{4.3.3}$$

For the integral is

$$-\int_0^{\pi/\lambda} H(t)\frac{d}{dt}\left(\frac{\sin \lambda t}{t}\right)dt = \int_0^{\pi/\lambda} o(t)\,O\left(\frac{\lambda}{t}\right)dt = o(1).$$

Plainly the same argument will prove that, if a is any positive constant, then

$$\int_{\tau_1}^{\tau_2} h(t)\frac{\sin \lambda t}{t}\,dt = o(1) \tag{4.3.4}$$

uniformly for $0 \leqq \tau_1 < \tau_2 \leqq a/\lambda$.

If now we take $h = g_c$ in (4.3.3), and combine the result with Theorem 51, we obtain

Theorem 53. *Suppose that $f(t)$ satisfies l_c, for $t = \theta$, and that*

(4.3.5) $$I(\delta, c, \lambda) = \int_{\pi/\lambda}^{\delta} g_c(t) \frac{\sin \lambda t}{t} dt.$$

Then a sufficient condition that $s_n(\theta) \to c$ is that there should be an $\eta(\epsilon)$ and a $\Lambda(\epsilon)$ such that $| I(\eta, c, \lambda) | < \epsilon$ for $\lambda \geqq \Lambda$. If l_c and the condition on I are satisfied uniformly in a set E in which $c = c(\theta)$ is bounded, then $s_n(\theta) \to c(\theta)$ uniformly in E.

For then

$$| J(\delta, c, \lambda) | < \epsilon + \left| \int_0^{\pi/\lambda} g_c \frac{\sin \lambda t}{t} dt \right| + \left| \int_{\eta}^{\delta} g_c \frac{\sin \lambda t}{t} dt \right|,$$

and $$\overline{\lim} \, | J(\delta, c, \lambda) | \leqq \epsilon,$$

by (4.3.3) and Theorem 31.

4.4. Dini's test. The simplest criterion for the convergence of a F.s. is due to Dini.

Theorem 54. *If $t^{-1} g_c(t)$ is L, for a certain c, then $s_n(\theta) \to c$. If in addition $t = \theta$ is a point of continuity or jump, then c is necessarily $f(\theta)$ or $\frac{1}{2}\{f(\theta+0)+f(\theta-0)\}$.*

This is an immediate corollary of Theorem 51, since $| J | < \epsilon$ by choice of η. It may also be deduced directly from the formula (4.2.2).

As a more special corollary, we have

Theorem 55. *If*

(4.4.1) $$f(\theta + t) - f(\theta) = O(| t |^{\alpha}),$$

where $\alpha > 0$, then the Fourier series of $f(t)$ converges to $f(\theta)$ for $t = \theta$. In particular this is true if $f'(\theta)$ exists (and is finite). If $f(t)$ satisfies (4.4.1) uniformly in a closed interval of θ, then the series converges uniformly in that interval.

4.5. Functions of bounded variation: Jordan's test. Our next group of theorems concerns functions of V, and we need a preliminary inequality. We suppose that h is a function of V, that $h(+0) = 0$, and that $V_h(t_1, t_2)$ is the variation of h in the open interval (t_1, t_2). Then

(4.5.1) $$\left| \int_{\tau}^{\delta} h(t) \frac{\sin \lambda t}{t} dt \right| \leqq H V_h(0, \delta) \quad (0 \leqq \tau \leqq \delta).$$

We can write $h = h^+ - h^-$, where h^+ and h^- are the positive and negative variations of h in $(0, t)$: h^+ and h^- are non-negative and increasing, $h^+(+0) = h^-(+0) = 0$, and $h^+ + h^- = V_h(0, t)$. Then

$$\left| \int_\tau^\delta h^+ \frac{\sin \lambda t}{t} \, dt \right| = \left| h^+(\delta - 0) \int_{\tau'}^\delta \frac{\sin \lambda t}{t} \, dt \right| = \left| h^+(\delta - 0) \int_{\tau'\lambda}^{\delta\lambda} \frac{\sin u}{u} \, du \right|,$$

where $\tau < \tau' < \delta$; and this does not exceed $HV_h(0, \delta)$. The same argument applies to h^-, and (4.5.1) follows.

Theorem 56. *If $\phi(t)$ is V in an interval $(0, \delta)$, then $s_n(\theta) \to \phi(+0)$. In particular, if $f(t)$ is V in an interval including θ, then the Fourier series of $f(t)$, for $t = \theta$, converges to $\frac{1}{2}\{f(\theta + 0) + f(\theta - 0)\}$.*

This is Jordan's test. It follows immediately from Theorem 51 and (4.5.1), since $|J(\eta, c, \lambda)| \leqq HV_h(0, \eta) < \epsilon$ when $h = g_c$, $c = \phi(+0)$, and η is sufficiently small.

The criteria for uniform or bounded convergence are more important here, and we state them separately.

Theorem 57. *If f is V in (a, b), then its Fourier series converges boundedly to $\frac{1}{2}\{f(\theta + 0) + f(\theta - 0)\}$ in any closed interval $\langle a', b' \rangle$ interior to (a, b). If also f is continuous in (a, b), then the series converges to $f(\theta)$ uniformly in $\langle a', b' \rangle$.*

If θ is in $\langle a', b' \rangle$, then $\theta + t$ is in (a, b) for $0 \leqq t \leqq \delta$ and sufficiently small δ, and $V_h(0, \delta) \leqq V_f(a, b)$ when $h = g_c$ and $c = \frac{1}{2}\{f(\theta + 0) + f(\theta - 0)\}$. The first clause of the theorem then follows from (4.5.1) and Theorem 50.

If f is also continuous, we take $h = g_c$ and $c = f(\theta)$. We can choose a $\delta(\epsilon)$, independent of θ, such that $V_h(0, \delta) < \epsilon$ for all θ of $\langle a', b' \rangle$; and this proves the second clause of the theorem. The theorem includes Theorem 45.

It should be observed that there is a fundamental difference between the tests of this section and those of § 4.4. Dini's test is strictly a *point test*, which tells us nothing about the convergence of the series except at the one point $t = \theta$. But if f is V in an interval round $t = \theta$, it is V in an interval round any neighbouring point, and the series is boundedly convergent in an interval round $t = \theta$.

If ξ is an *isolated* discontinuity of a function f of V, then the F.s. of the g of (3.12.4) is uniformly convergent near ξ, so that, after § 3.12, the F.s. of f shows the Gibbs phenomenon at ξ.

Theorem 57 gives us an alternative proof of Theorems 43 and 44 (and so of the completeness of the trigonometrical systems). For $F(\theta)$ is absolutely continuous, *a fortiori* continuous and V, and so its F.s.

$$\tfrac{1}{2}A_0 - \Sigma n^{-1}(b_n \cos n\theta - a_n \sin n\theta)$$

is uniformly convergent. Putting $\theta = 0$, we see that $A_0 = 2\Sigma n^{-1} b_n$.

4.6. Lebesgue's test.

Our final test is a refinement, due to Gergen, of one given originally by Lebesgue.

Theorem 58. *If f satisfies l_c for $t = \theta$, and*

$$\int_h^\delta \frac{|\phi(t+h) - \phi(t)|}{t}\, dt \to 0, \tag{4.6.1}$$

for some fixed δ, when $h \to +0$, then $s_n(\theta) \to c$.

It should be observed that ϕ may be replaced by g_c in (4.6.1), since $g_c(t+h) - g_c(t)$ is independent of c.

It is sufficient, after Theorem 53, to show that

$$I = I(\delta, c, \lambda) = \int_h^\delta \frac{g(t)}{t} \sin \lambda t\, dt = o(1),$$

where $h = \pi/\lambda$. We can change the upper limit in I to $\delta + h$, and the lower limit to $2h$, with error $o(1)$; the first change being trivial and the second justified by (4.3.4). If we do this, and write $t+h$ for t, we obtain

$$I = -\int_h^\delta \frac{g(t+h)}{t+h} \sin \lambda t\, dt + o(1).$$

It follows, adding the two expressions for I, that

$$I = \frac{1}{2}\int_h^\delta \left\{\frac{g(t)}{t} - \frac{g(t+h)}{t+h}\right\} \sin \lambda t\, dt + o(1) \tag{4.6.2}$$

$$= \frac{1}{2}\int_h^\delta \frac{g(t) - g(t+h)}{t+h} \sin \lambda t\, dt + \tfrac{1}{2}h \int_h^\delta \frac{g(t)}{t(t+h)} \sin \lambda t\, dt + o(1).$$

The first term here does not exceed

$$\frac{1}{2}\int_h^\delta \frac{|g(t+h) - g(t)|}{t}\, dt = o(1),$$

by (4.6.1), and it is therefore sufficient to prove that

$$P = \int_h^\delta \frac{g(t)}{t(t+h)} \sin \lambda t\, dt = o(\lambda). \tag{4.6.3}$$

Now

$$P = \left(\int_h^{2h} + \int_{2h}^{\delta}\right) \frac{g(t)}{t(t+h)} \sin \lambda t \, dt = P_1 + P_2, \tag{4.6.4}$$

say. Here
$$P_1 = \frac{1}{2h} \int_h^{\tau} \frac{g(t)}{t} \sin \lambda t \, dt,$$

where $h < \tau < 2h$, and this is $o(h^{-1}) = o(\lambda)$, by (4.3.4). Also

$$P_2 = \int_{2h}^{\delta} \frac{g(t)}{t(t+h)} \sin \lambda t \, dt = -\int_h^{\delta-h} \frac{g(t+h)}{(t+h)(t+2h)} \sin \lambda t \, dt,$$

and we may replace $\delta - h$ here by δ with error $o(1)$. If we do this, substitute in (4.6.4), remember that $P_1 = o(\lambda)$, and then take the mean of the two forms of P, we obtain

$$P = \frac{1}{2} \int_h^{\delta} \left\{\frac{g(t)}{t(t+h)} - \frac{g(t+h)}{(t+h)(t+2h)}\right\} \sin \lambda t \, dt + o(\lambda) = P_3 + hP_4 + o(\lambda),$$

where

$$P_3 = \frac{1}{2} \int_h^{\delta} \frac{g(t) - g(t+h)}{(t+h)(t+2h)} \sin \lambda t \, dt, \quad P_4 = \int_h^{\delta} \frac{g(t)}{t(t+h)(t+2h)} \sin \lambda t \, dt.$$

Here
$$| P_3 | \leqq \frac{1}{6h} \int_h^{\delta} \frac{| g(t+h) - g(t) |}{t} dt = o(\lambda),$$

by (4.6.1). It is therefore enough to prove that $P_4 = o(\lambda^2)$.

Now, if $G(t)$ is defined as $H(t)$ was defined in § 4.3,

$$\int_h^{\delta} \frac{g(t)}{t(t+h)(t+2h)} \sin \lambda t \, dt$$
$$= \frac{G(\delta) \sin \lambda\delta}{\delta(\delta+h)(\delta+2h)} - \int_h^{\delta} G(t) \frac{d}{dt}\left\{\frac{\sin \lambda t}{t(t+h)(t+2h)}\right\} dt,$$

and the term integrated out is $O(1)$. Also $G(t) = o(t)$, since f satisfies l_c, and

$$\left| \frac{d}{dt}\left\{\frac{\sin \lambda t}{t(t+h)(t+2h)}\right\} \right| < H\left(\frac{\lambda}{t^3} + \frac{1}{t^4}\right).$$

Hence the last integral is

$$o\left(\lambda \int_h^{\delta} \frac{dt}{t^2}\right) + o\left(\int_h^{\delta} \frac{dt}{t^3}\right) = o(\lambda^2);$$

and this completes the proof of the theorem.

The proof of Theorem 58 is sophisticated. It is much simplified if we take

$$\int_h^\delta \left| \frac{g_c(t+h)}{t+h} - \frac{g_c(t)}{t} \right| dt \to 0 \tag{4.6.5}$$

as hypothesis instead of (4.6.1), since then it follows at once from the first line of (4.6.2) that $I \to 0$. This hypothesis can be satisfied for only one value of c, and it can be shown that it *implies* L_c and *a fortiori* l_c, so that (4.6.5) is in itself a sufficient condition for convergence.

4.7. Another test for uniform convergence. Lebesgue's test is the most comprehensive of the convergence tests at present known; it may be shown, for example, that it includes both Dini's and Jordan's. Its defect is that, while very comprehensive (and in this respect, perhaps, approaching the 'ultimate truth'), it is not very illuminating: it does not correspond, in any obvious way, with any of the most natural characteristics of f.

There is an interesting test for uniform convergence which is easily deduced from Theorem 58.

Theorem 59. *If*

$$| f(\theta+h) - f(\theta) | = o\left(\log \frac{1}{|h|}\right)^{-1} \tag{4.7.1}$$

uniformly in an open interval, then the Fourier series converges uniformly to $f(\theta)$ *in any closed interval interior to that interval.*

For f satisfies l_f uniformly, and (supposing h positive)

$$\int_h^\delta \frac{| g_c(t+h) - g_c(t) |}{t} dt = o\left\{\left(\log \frac{1}{h}\right)^{-1} \int_h^\delta \frac{dt}{t}\right\} = o(1),$$

uniformly in any closed interior interval. The theorem is particularly interesting because (4.7.1), for a special θ, does *not* imply convergence for that θ.

4.8. The conjugate series. The theorems proved in the preceding sections have analogues for the 'conjugate series' $\tilde{T}(f)$ defined in (1.2.9) and § 1.3. It is to be remembered that, as we saw in § 3.10, the c.s. is not necessarily itself a F.s., and that, even if it is one, we want criteria for its convergence stated in terms of f.

We denote the partial sum of $\tilde{T}(f)$ by $\tilde{s}_n(\theta)$ or $\tilde{s}_n(\theta, f)$. We write

$$\psi(t) = \psi(\theta, t) = f(\theta+t) - f(\theta-t). \tag{4.8.1}$$

The F.c. of $\psi(t)$ are 0 and $2B_n(\theta)$, by Theorem 26; and the c.s. of $f(t)$, for $t = \theta$, is the c.s. of $\frac{1}{2}\psi(t)$, a cosine series in t, for $t = 0$.

It is not immediately obvious what is the 'natural' sum for $\tilde{T}(f)$. We shall find that it is

$$(4.8.2) \quad \tilde{f}(\theta) = \frac{1}{2\pi}\int_0^\pi \psi(\theta, t) \cot \tfrac{1}{2}t\, dt = \lim_{\epsilon\to 0} \frac{1}{2\pi}\int_\epsilon^\pi \psi(\theta, t) \cot \tfrac{1}{2}t\, dt,$$

the integral being usually a 'Cauchy integral' at the origin. The existence of the integral in this sense is a problem in itself, to which we shall return in Ch. VI. In this chapter we are concerned not so much with $\tilde{f}(\theta)$, or the partial sum $\tilde{s}_n(\theta)$, as with the difference

$$(4.8.3) \quad \tilde{d}_n(\theta) = \tilde{s}_n(\theta) - \tilde{f}_n(\theta) = \tilde{s}_n(\theta) - \frac{1}{2\pi}\int_{\pi/n}^\pi \psi(\theta, t) \cot \tfrac{1}{2}t\, dt.$$

We find that to each of our tests for the convergence of $s_n(\theta)$ corresponds a test for the convergence of $\tilde{d}_n(\theta)$: if we add, as an additional hypothesis, the existence of $\tilde{f}(\theta)$, then $\tilde{f}_n(\theta)$ will tend to $\tilde{f}(\theta)$, and we shall obtain a test for the convergence of $\tilde{T}(f)$.

It is plain that $\tilde{f}(\theta)$ cannot exist at a point of jump of $f(\theta)$. We shall find that the 'natural' limit for $\tilde{d}_n(\theta)$, at such a point, is βd, where

$$(4.8.4) \quad \beta = (\gamma + \log \pi)/\pi,$$

γ is Euler's constant, and $d = \psi(+0)$ is the jump. Consider, for example, the series

$$(4.8.5) \quad T(\theta) = \sin\theta + \tfrac{1}{3}\sin 3\theta + \ldots, \quad \tilde{T}(\theta) = \cos\theta + \tfrac{1}{3}\cos 3\theta + \ldots,$$

for $\theta = 0$. In this case $f(\theta) = \frac{1}{4}\pi \operatorname{sgn}\theta$ for $-\pi < \theta < \pi$, by (3.7.4), so that $\phi(t) = 0$ and $\psi(t) = \frac{1}{2}\pi$ for $0 < t < \pi$. Plainly $T(0)$ converges to $\phi(+0) = 0$. Also

$$(4.8.6) \quad \tilde{s}_n(0) = \sum_{2m+1 \le n} \frac{1}{2m+1} = \tfrac{1}{2}\log n + \tfrac{1}{2}\log 2 + \tfrac{1}{2}\gamma + o(1)$$

and $$\tilde{f}_n(0) = \frac{1}{4}\int_{\pi/n}^{\pi} \cot\tfrac{1}{2}\theta\, d\theta = -\tfrac{1}{2}\log\sin\frac{\pi}{2n} = \tfrac{1}{2}\log n + \tfrac{1}{2}\log\frac{2}{\pi} + o(1),$$

so that $$\tilde{d}_n(0) \to \tfrac{1}{2}(\gamma + \log\pi) = \beta d.$$

There is no particular significance in the constant β, whose value is determined by our choice of π/n as the lower limit of the integral in (4.8.3). If we chose a/n, β would be $(\gamma + \log a)/\pi$.

4.9. The convergence problem for the conjugate series.

The discussion of the c.s. is similar to that of the F.s., but is a little complicated by the distinction between $\tilde{s}_n(\theta)$ and $\tilde{d}_n(\theta)$ and the divergence of $\tilde{s}_n(\theta)$ at a point of jump. Here

$$(4.9.1) \qquad \tilde{s}_n(\theta) = \sum_1^n B_m(\theta) = \frac{1}{\pi}\int_{-\pi}^{\pi} f(t) \sum_1^n \sin m(t-\theta)\, dt$$

$$= \frac{1}{\pi}\int_{-\pi}^{\pi} f(\theta+t) \sum_1^n \sin mt\, dt = \frac{1}{2\pi}\int_{-\pi}^{\pi} f(\theta+t) \frac{\cos\frac{1}{2}t - \cos(n+\frac{1}{2})t}{\sin\frac{1}{2}t}\, dt$$

$$= \frac{1}{2\pi}\int_0^{\pi} \psi(\theta,t) \frac{\cos\frac{1}{2}t - \cos(n+\frac{1}{2})t}{\sin\frac{1}{2}t}\, dt$$

$$= \frac{1}{2\pi}\int_0^{\pi} \psi(\theta,t)\cot\tfrac{1}{2}t(1-\cos nt)\, dt + \frac{1}{2\pi}\int_0^{\pi} \psi(\theta,t)\sin nt\, dt.$$

The last term is $o(1)$. In particular, if $T(\theta)$ is the series in (4.8.5), and $\theta = 0$, then $\psi = \frac{1}{2}\pi$, and $\tilde{s}_n(0)$ is given by (4.8.6); so that

$$(4.9.2) \qquad \tfrac{1}{2}\log 2n + \tfrac{1}{2}\gamma = \frac{1}{4}\int_0^{\pi} \cot\tfrac{1}{2}t(1-\cos nt)\, dt + o(1).$$

Multiplying (4.9.2) by $2d/\pi$, where $d = d(\theta)$, subtracting the result from (4.9.1), and writing

$$(4.9.3) \qquad \tilde{g}_d(t) = \psi(\theta,t) - d = f(\theta+t) - f(\theta-t) - d,$$

we obtain

(4.9.4)

$$\tilde{s}_n(\theta) - \frac{d}{\pi}(\log 2n + \gamma) = \frac{1}{2\pi}\int_0^{\pi} \tilde{g}_d(t)\cot\tfrac{1}{2}t(1-\cos nt)\, dt + o(1).$$

Hence, if $\tilde{d}_n(\theta)$ is defined by (4.8.3), we have

$$\tilde{d}_n(\theta) - \frac{d}{\pi}(\log 2n + \gamma)$$

$$= \frac{1}{2\pi}\int_0^{\pi/n} \tilde{g}\cot\tfrac{1}{2}t(1-\cos nt)\, dt - \frac{1}{2\pi}\int_{\pi/n}^{\pi} \tilde{g}\cot\tfrac{1}{2}t\cos nt\, dt - R + o(1),$$

where

$$R = \frac{d}{2\pi}\int_{\pi/n}^{\pi} \cot\tfrac{1}{2}t\, dt = -\frac{d}{\pi}\log\sin\frac{\pi}{2n} = \frac{d}{\pi}(\log 2n - \log\pi) + o(1);$$

and therefore

$$(4.9.5) \qquad \tilde{d}_n(\theta) - \beta d = \frac{1}{2\pi}\int_0^{\pi/n} \tilde{g}\cot\tfrac{1}{2}t(1-\cos nt)\, dt$$

$$- \frac{1}{2\pi}\int_{\pi/n}^{\pi} \tilde{g}\cot\tfrac{1}{2}t\cos nt\, dt + o(1).$$

We can now argue with (4.9.5) substantially as we argued with

(4.2.2) in § 4.2, replacing π by δ, $\frac{1}{2}\cot\frac{1}{2}t$ by $1/t$, and n by a continuous λ. We thus obtain

Theorem 60. *In order that*

$$\tilde{d}_n(\theta) \to \beta d, \tag{4.9.6}$$

it is necessary and sufficient that

$$\tilde{J} = \tilde{J}(\delta, d, \lambda) = \int_0^{\pi/\lambda} \tilde{g}_d(t) \frac{1-\cos\lambda t}{t}\,dt - \int_{\pi/\lambda}^{\delta} \tilde{g}_d(t) \frac{\cos\lambda t}{t}\,dt \to 0, \tag{4.9.7}$$

for a fixed positive δ, when $\lambda \to \infty$. It is sufficient that there should be an $\eta = \eta(\epsilon)$ and a $\Lambda = \Lambda(\epsilon)$, corresponding to every positive ϵ, such that $|\tilde{J}(\eta, d, \lambda)| < \epsilon$ for $\lambda \geqq \Lambda$.

We could add a supplementary clause concerning the uniform convergence, or boundedness, of $\tilde{d}_n(\theta)$, in sets in which $d = d(\theta)$ is bounded; but we leave questions of uniformity concerning the c.s. to the reader.

The convergence of $\tilde{d}_n$, like that of s_n, depends only on the values of f near θ, and so does the convergence of $\tilde{s}_n$; but the value of $\tilde{f}$, and the sum of $\tilde{T}(f)$, naturally depend on *all* the values of f.

4.10. Criteria for the convergence of the conjugate series. There are further developments corresponding to those of § 4.3, and criteria corresponding to those of §§ 4.4–5.

We define the conditions $\tilde{l}_d$ and $\tilde{L}_d$ as we defined l_c and L_c in § 4.3, but with $h = \tilde{g}_d$. If h satisfies (4.3.1), then

$$\int_0^{\pi/\lambda} h(t) \frac{1-\cos\lambda t}{t}\,dt = o(1): \tag{4.10.1}$$

the proof is similar to that of (4.3.3). We thus obtain

Theorem 61. *If $f(t)$ satisfies $\tilde{l}_d$, for $t = \theta$, and*

$$\tilde{I}(\delta, d, \lambda) = \int_{\pi/\lambda}^{\delta} \tilde{g}_d(t) \frac{\cos\lambda t}{t}\,dt, \tag{4.10.2}$$

then a sufficient condition that $\tilde{d}_n(\theta) \to \beta d$ is that there should be an $\eta(\epsilon)$ and a $\Lambda(\epsilon)$ such that $|\tilde{I}| < \epsilon$ for $\lambda \geqq \Lambda$.

This is the analogue of Theorem 53 (apart from the clause concerning uniformity). The analogues of Theorems 54 and 56 are

Theorem 62. *If $t^{-1}\tilde{g}_d(t)$ is L, for a certain d, then*

$$\tilde{d}_n(\theta) \to \beta d, \quad \tilde{s}_n(\theta) - \frac{d}{\pi}(\log 2n + \gamma) \to \frac{1}{2\pi}\int_0^{\pi} \tilde{g}_d(t)\cot\tfrac{1}{2}t\,dt.$$

If $t = \theta$ is a point of jump, d must be $f(\theta+0)-f(\theta-0)$. If it is a point of continuity, d must be 0; and then $\tilde{d}_n(\theta)\to 0$ and $\tilde{s}_n(\theta)\to\tilde{f}(\theta)$.

Theorem 63. *If $\psi(t)$ is V in an interval $(0, \delta)$, then*

$$\tilde{d}_n(\theta)\to\beta\psi(+0).$$

If also $\tilde{f}(\theta)$ exists, then $\psi(+0) = 0$ and $\tilde{T}(f)$ converges to $\tilde{f}(\theta)$.

Theorem 62 follows from Theorem 61, or may be deduced directly from (4.9.5). The proof of Theorem 63 is similar to that of Theorem 56, but we need an additional inequality, viz.

$$\left|\int_\tau^\delta h(t)\frac{\cos\lambda t}{t}\,dt\right| \leqq HV_h(0,\delta) \quad \left(0<\frac{1}{\lambda}\leqq\tau\leqq\delta\right).$$

There is also an analogue of Theorem 58.

We add one remark which will be useful in Ch. v. If

$$\mathbf{H}(t) = \int_0^t \frac{h(u)}{u}\,du$$

exists, as a Cauchy integral down to 0, then

$$H(t) = \int_0^t h(u)\,du = \int_0^t u\mathbf{H}'(u)\,du = t\mathbf{H}(t) - \int_0^t \mathbf{H}(u)\,du = o(t),$$

so that $h(t)$ satisfies (4.3.1). If $\mathbf{H}(t)$ exists as a Lebesgue integral, then $h(t)$ satisfies (4.3.2). In these cases (4.10.1) may be replaced by

$$\int_0^{\pi/\lambda} h(t)\frac{\cos\lambda t}{t}\,dt = o(1). \tag{4.10.3}$$

4.11. The order of magnitude of $s_n(\theta)$ and $\tilde{s}_n(\theta)$. The conditions L_c and $\tilde{L}_d$ are satisfied p.p. when $c = f(\theta)$ and $d = 0$; and it is important to know how much can be said about $s_n(\theta)$ and $\tilde{s}_n(\theta)$ when we assume these conditions only.

Theorem 64. *If f satisfies L_c, then*

$$s_n(\theta) = o(\log n). \tag{4.11.1}$$

If f satisfies $\tilde{L}_d$, then

$$\tilde{s}_n(\theta) \sim (d/\pi)\log n. \tag{4.11.2}$$

In particular, both $s_n(\theta)$ and $\tilde{s}_n(\theta)$ are $o(\log n)$ for almost all θ.

(1) Take $h = g_c$, $\lambda = n$. Then

$$\left|\int_{\pi/n}^\pi h(t)\frac{\sin nt}{t}\,dt\right| \leqq \int_{\pi/n}^\pi \frac{|h(t)|}{t}\,dt \leqq \frac{H^*(\pi)}{\pi} + \int_{\pi/n}^\pi \frac{H^*}{t^2}\,dt$$

$$\leqq O(1) + \int_{\pi/n}^\pi o\left(\frac{1}{t}\right)dt = o(\log n).$$

(2) Take $h = \tilde{g}_d$. Then the same argument, with $\cos nt$ for $\sin nt$, shows that $\tilde{d}_n(\theta) = o(\log n)$. Also

$$\frac{1}{2\pi}\int_{\pi/n}^{\pi} \psi \cot \tfrac{1}{2}t\,dt = \frac{1}{2\pi}\int_{\pi/n}^{\pi} \tilde{g}_d \cot \tfrac{1}{2}t\,dt + \frac{d}{2\pi}\int_{\pi/n}^{\pi} \cot \tfrac{1}{2}t\,dt.$$

A similar argument shows that the first integral is $o(\log n)$; and the second is asymptotic to $(d/\pi)\log n$.

Corollaries of (4.11.2) are (1) that the c.s. diverges to $\pm\infty$ at a point of jump, and (2) that a function whose F.c. are $o(n^{-1})$ cannot have a jump.

4.12. Divergence at a point of continuity. It was for long uncertain whether a continuity condition of the type of l_c or L_c might not be by itself sufficient for the convergence of the F.s. (in which case the theorems which we have proved would lose nearly all their significance).

Theorem 65. *There are functions whose Fourier series diverge at points of continuity.*

We suppose that N_r is an odd integer, at least 3, and

(4.12.1) $n_0 = 1,\quad n_r = N_1 N_2 \ldots N_r,\quad a_r > 0,\quad \Sigma a_r < \infty,\quad a_r \log N_r \to \infty.$

All these conditions are satisfied, for example, if $a_r = r^{-2}$ and $n_r = 3^{r^4}$. We define $f(t)$ in $\langle 0, \pi\rangle$ by

$$f(0) = 0,\quad f(t) = a_r \sin n_r t \quad (\pi/n_r \leqq t \leqq \pi/n_{r-1}),$$

and by evenness and periodicity elsewhere. Then f is continuous in $\langle -\pi, \pi\rangle$, and V in any sub-interval which does not include 0, so that its F.s. is uniformly convergent in any such sub-interval. We prove that it diverges at 0.

It is enough to show that

$$J(k) = \int_0^{\pi} f(t)\frac{\sin n_k t}{t}\,dt \to \infty$$

when $k \to \infty$. Now

$$J(k) = \sum_1^{\infty} a_r \int_{\pi/n_r}^{\pi/n_{r-1}} \frac{\sin n_k t \sin n_r t}{t}\,dt = \sum_1^{\infty} a_r j_r(k) = a_k j_k(k) + S(k),$$

say, where $S(k)$ contains the terms for which $r \neq k$. If $r \neq k$, then

$$j_r(k) = \frac{1}{2}\int_{\pi/n_r}^{\pi/n_{r-1}} \frac{\cos|n_k - n_r|t - \cos(n_k + n_r)t}{t}\,dt = \frac{1}{2}\sum_{j=1}^{4} \epsilon_j \int_1^{U_j} \frac{\cos u}{u}\,du,$$

where ϵ_j is ± 1 and U_j is one of

$$\frac{\pi \mid n_k - n_r \mid}{n_r}, \quad \frac{\pi \mid n_k - n_r \mid}{n_{r-1}}, \quad \frac{\pi(n_k + n_r)}{n_r}, \quad \frac{\pi(n_k + n_r)}{n_{r-1}}.$$

Since $n_r/n_{r-1} \geqq 3$ for all r, none of these numbers is less than $\frac{2}{3}\pi$. Hence $j_r(k)$ is uniformly bounded, and therefore $S(k)$ is bounded.

On the other hand

$$j_k(k) = \frac{1}{2}\int_{\pi/n_k}^{\pi/n_{k-1}} \frac{1 - \cos 2n_k t}{t} dt = \tfrac{1}{2}\log N_k - \frac{1}{2}\int_{2\pi}^{2\pi N_k} \frac{\cos u}{u} du,$$

and the last term is bounded. It now follows from (4.12.1) that $a_k j_k(k) \to \infty$ and $J(k) \to \infty$.

4.13. The Lebesgue functions of a normal orthogonal system. Theorem 65 is a special case of a much more general theorem.

Suppose that (ϕ_n) is a real n.o.s. for (a, b), and that $f \sim (c_n)$ for this system. The nth Fourier polynomial of f is

$$(4.13.1) \quad s_n(x, f) = \int_a^b f(y) \sum_1^n \phi_m(x)\, \phi_m(y)\, dy = \int_a^b f(y)\, \Phi_n(x, y)\, dy,$$

say. If $|f| \leqq 1$, then

$$(4.13.2) \qquad |s_n(x, f)| \leqq \int_a^b |\Phi_n(x, y)|\, dy = \rho_n(x).$$

We call $\rho_n(x)$ the nth *Lebesgue function* of (ϕ_n).

Theorem 66. *If* $\overline{\lim}\, \rho_n(x) = \infty$ *for some* x, *then there is a continuous function* f *whose Fourier series diverges for that* x.

If $f(y) = \operatorname{sgn} \Phi_n(x, y)$, then $s_n(x, f) = \rho_n(x)$, by (4.13.1) and (4.13.2). This f is discontinuous, but it is plain that, by taking a suitable continuous approximation to f, we can find a continuous f_n such that

$$(4.13.3) \qquad |f_n(y)| \leqq 1, \quad s_n(x, f_n) \geqq \tfrac{1}{2}\rho_n(x).$$

We do this for each n. If the F.s. of any f_n diverges at x, the theorem is proved. We may therefore suppose that the F.s. of each f_n converges, say to γ_n.

We now define $F(y)$ by

$$(4.13.4) \qquad F(y) = \Sigma \alpha_k f_{n_k}(y),$$

where

$$(4.13.5) \qquad \alpha_k > 0, \quad \sum_1^\infty \alpha_k < \infty, \quad \sum_{k+1}^\infty \alpha_p \leqq \tfrac{1}{6}\alpha_k;$$

for example, we may take $\alpha_k = 7^{-k}$. Then F is continuous, and its F.s. is obtained by formal addition of the F.s. of the terms of (4.13.4).

We choose the n_k by induction, as follows. Given $n_1, n_2, \ldots, n_{k-1}$, we can choose n_k so that

$$\alpha_k \rho_{n_k} \to \infty, \quad \sum_1^{k-1} \alpha_p \, |\gamma_{n_p}| \leqq \tfrac{1}{12}\alpha_k \rho_{n_k}; \tag{4.13.6}$$

and, since
$$s_n\left(x, \sum_1^{k-1} \alpha_p f_{n_p}\right) \to \sum_1^{k-1} \alpha_p \gamma_{n_p},$$

we may suppose n_k so large that

$$\left| s_{n_k}\left(x, \sum_1^{k-1} \alpha_p f_{n_p}\right)\right| \leqq 2 \sum_1^{k-1} \alpha_p \, |\gamma_{n_p}| \leqq \tfrac{1}{6}\alpha_k \rho_{n_k}. \tag{4.13.7}$$

Also

$$\left| s_{n_k}\left(x, \sum_{k+1}^{\infty} \alpha_p f_{n_p}\right)\right| \leqq \rho_{n_k} \sum_{k+1}^{\infty} \alpha_p \leqq \tfrac{1}{6}\alpha_k \rho_{n_k}, \tag{4.13.8}$$

by (4.13.2) and (4.13.5). Finally, it follows from (4.13.3), (4.13.7), and (4.13.8) that

$$s_{n_k}(x, F) \geqq s_{n_k}(x, \alpha_k f_{n_k}) - \tfrac{1}{6}\alpha_k \rho_{n_k} - \tfrac{1}{6}\alpha_k \rho_{n_k} \geqq \tfrac{1}{6}\alpha_k \rho_{n_k} \to \infty,$$

so that the F.s. of F diverges at x.

4.14. The Lebesgue constants of the trigonometrical system (T).

For the real trigonometrical system

$$\rho_n(x) = \frac{1}{2\pi}\int_0^{2\pi} \left| \frac{\sin(n+\frac{1}{2})(t-x)}{\sin\frac{1}{2}(t-x)} \right| dt = \frac{1}{2\pi}\int_0^{2\pi} \left| \frac{\sin(n+\frac{1}{2})t}{\sin\frac{1}{2}t} \right| dt = \rho_n$$

is independent of x. We call ρ_n the nth *Lebesgue constant* of (T).

Theorem 67: $\rho_n \sim (4/\pi^2)\log n$.

In fact

$$\rho_n = \frac{1}{\pi}\int_0^{\pi} |\sin nt \cot\tfrac{1}{2}t + \cos nt| \, dt$$

$$= \frac{1}{\pi}\int_0^{\pi} |\sin nt \cot\tfrac{1}{2}t| \, dt + O(1) = \sigma_n + O(1),$$

where
$$\sigma_n = \frac{2}{\pi}\int_0^{\pi} \frac{|\sin nt|}{t} dt = \frac{2}{\pi}\int_0^{n\pi} \frac{|\sin u|}{u} du.$$

But
$$\sigma_{n+1} - \sigma_n = \frac{2}{\pi}\int_{n\pi}^{(n+1)\pi} \frac{|\sin u|}{u} du \sim \frac{2}{\pi^2 n}\int_0^{\pi} \sin u \, du = \frac{4}{\pi^2 n}.$$

Hence
$$\rho_n \sim \sigma_n \sim (4/\pi^2)\log n.$$

Since $\rho_n \to \infty$, Theorems 66 and 67 give an alternative proof of Theorem 65.

V. SUMMABILITY OF FOURIER SERIES

5.1. Introduction. The theorems of Ch. IV establish the convergence of the F.s. of various natural and extensive classes of functions, but they have serious limitations. In the first place, as we showed in § 4.12, a F.s. is not necessarily convergent at a point of continuity of its generating function. Secondly (though this we have not yet proved) it is not necessarily convergent p.p. It is therefore natural to ask whether there are not methods of summation, among the many used in the modern theory of divergent series, which sum F.s. more effectively.

The first and most important step in this direction was taken by Fejér in 1904, when he showed that

$$\sigma_n(\theta, f) = \frac{1}{n+1}\{s_0(\theta, f) + s_1(\theta, f) + \dots + s_n(\theta, f)\}$$

tends to $f(\theta)$ at any point of continuity of $f(t)$. A little later Lebesgue proved that $\sigma_n(\theta, f) \to f(\theta)$ p.p. These results showed that this method of summation by arithmetic means, now usually called the $(C, 1)$ method, succeeds at the two fundamental points where ordinary convergence fails, and marked a turning-point in the development of the theory of F.s. Poisson, many years before, had, in effect, summed F.s. by what is now called the 'Poisson' or 'Abel' or A method; but his point of view was very different, and his ideas had not the precision required in the modern theory.

A method of summation which sums the F.s. of any integrable $f(\theta)$, to $f(\theta)$, p.p. may be called 'Fourier-effective'. Then, if $f_1(\theta)$ is the sum, $f_1 \equiv f$, and the F.s. of f_1 is the same as that of f. Otherwise it might be possible to define a sequence of different functions, f_1 the sum of the F.s. of f, f_2 the sum of the F.s. of f_1, and so on. For a 'Fourier-effective' method, $f_1, f_2, \dots$ are the same.

5.2. Linear and regular methods of summation. The most important methods of summation are *linear*. Toeplitz was the first to consider such methods systematically, and we shall call them T-methods. Given a matrix

$$((\alpha_{m,n})) \quad (m = 0, 1, 2, \dots;\ n = 0, 1, 2, \dots),$$

we define the *transforms* τ_m of the sequence (s_n) by

(5.2.1) $$\tau_m = \sum_{n=0}^{\infty} \alpha_{m,n} s_n.$$

We assume that all these series are convergent. If

(5.2.2) $$\tau_m \to s$$

when $m \to \infty$, we say that s_n *has the* T-*limit* s or *converges* (T) *to* s, and write

(5.2.3) $$s_n \to s \text{ (T)}.$$

We shall also say that the method T is *effective for* (s_n). It is plain that $s_n \to s$ (T) and $t_n \to t$ (T) imply $As_n + Bt_n \to As + Bt$ (T): this is a characteristic of a linear method. If $\alpha_{m,n} = 0$ unless $n = m$, and $\alpha_{m,m} = 1$, then $\tau_m = s_m$, and T-convergence is ordinary convergence.

We shall usually be concerned with sequences

(5.2.4) $$s_n = u_0 + u_1 + \ldots + u_n$$

defined by partial sums of infinite series. We then call T a method of summation of series: if s_n has the T-limit s, we say that

$$\Sigma u_n = u_0 + u_1 + \ldots$$

is *summable* (T) *to sum* s, and write

(5.2.5) $$u_0 + u_1 + u_2 + \ldots = s \text{ (T)}.$$

If T sums every convergent series to the right sum, i.e. if $s_n \to s$ implies $s_n \to s$ (T), then we say that T is a *regular* method. The conditions on the $\alpha_{m,n}$ which make T regular were found by Toeplitz, and we state his result as a formal theorem.

Theorem 68. *In order that* T *should be regular, it is necessary and sufficient that*

(5.2.6) $$R_m = \sum_n |\alpha_{m,n}| < R,$$

where R is independent of m;

(5.2.7) $$\alpha_{m,n} \to 0$$

when $m \to \infty$, for every n; and

(5.2.8) $$r_m = \sum_n \alpha_{m,n} \to 1.$$

The conditions are necessary in the sense that, if they are not satisfied, there are series convergent to s but not summable (T) to s. We add

(i) that the first two conditions are enough to ensure that $\tau_m \to 0$ whenever $s_n \to 0$;

(ii) that if $\alpha_{m,n} \geqq 0$ and s_n is real, then

$$\underline{\lim}\, s_n \leqq \underline{\lim}\, \tau_m \leqq \overline{\lim}\, \tau_m \leqq \overline{\lim}\, s_n;$$

(iii) that if also $r_m = 1$, then

$$\operatorname{Min} s_n \leqq \tau_m \leqq \operatorname{Max} s_n,$$

where 'Min' and 'Max' are the lower and upper bounds for all n;

(iv) that, if the conditions are satisfied uniformly in any additional parameter in $\alpha_{m,n}$, then the conclusions also are satisfied uniformly.

The transforms (5.2.1) depend on the integral parameter m. It is sometimes convenient to use transforms $\tau(r)$ depending on a continuous parameter r which tends to ∞, 0, 1 or some other limit. The difference is unimportant, since it is the same thing to say, for example, that $\tau(r) \to s$ when $r \to 1$, as to say that $\tau(r_m) \to s$ when $m \to \infty$ and (r_m) is an arbitrary sequence of values of r with limit 1. In particular Theorem 68 remains true with the obvious verbal changes.

5.3. The $(C, 1)$ and A methods. We shall usually be concerned with one or other of two standard methods.

(1) If $\quad \alpha_{m,n} = \dfrac{1}{m+1} \quad (n \leqq m), \qquad \alpha_{m,n} = 0 \quad (n > m),$

then
$$\tau_m = \frac{s_0 + s_1 + \ldots + s_m}{m+1} = \sigma_m.$$

We call this method the $(C, 1)$ method. Since $\alpha_{m,n} \geqq 0$ and $r_m = 1$, the method is regular.

(2) The A method depends on a continuous parameter r which tends to 1 through positive values less than 1. If

$$\alpha_{r,n} = \alpha_n(r) = (1-r)\, r^n,$$

then
$$\tau(r) = (1-r)\, \Sigma s_n r^n = \Sigma u_n r^n = \sigma(r)$$

whenever the first series is convergent. We can verify at once that the method satisfies the conditions of Theorem 68 (modified for the continuous parameter), and is therefore regular. This is the substance of Abel's theorem on the continuity of power-series.

A series summable $(C, 1)$ is summable (A) to the same sum; but the A method is much more powerful.

(3) We shall need the values of σ_m and $\sigma(r)$ for the special series

(C) $$\tfrac{1}{2}+\cos\theta+\cos 2\theta+\ldots$$

and

(S) $$\sin\theta+\sin 2\theta+\ldots.*$$

We find by straightforward calculation that

(5.3.1) $$\sigma_m = \frac{1}{m+1}\sum_0^m D_n(\theta) = \frac{1}{2(m+1)}\left\{\frac{\sin\frac{1}{2}(m+1)\theta}{\sin\frac{1}{2}\theta}\right\}^2 = F_m(\theta),$$

(5.3.2) $$\sigma(r) = \frac{1}{2}\,\frac{1-r^2}{1-2r\cos\theta+r^2} = P(r,\theta)$$

for C; and

(5.3.3)
$$\sigma_m = \frac{1}{m+1}\sum_1^m \tilde{D}_n(\theta) = \tfrac{1}{2}\cot\tfrac{1}{2}\theta - \frac{1}{m+1}\,\frac{\sin(m+1)\theta}{(2\sin\frac{1}{2}\theta)^2} = G_m(\theta),$$

(5.3.4) $$\sigma(r) = \frac{r\sin\theta}{1-2r\cos\theta+r^2} = Q(r,\theta)$$

for S.

It follows that either of the methods sums C to 0 and S to $\frac{1}{2}\cot\frac{1}{2}\theta$ for any θ which is not a multiple of 2π, and uniformly in any closed interval free from multiples of 2π. It is plain that

(5.3.5) $$F_m(\theta) \geqq 0, \quad P(r,\theta) \geqq 0,$$

(5.3.6) $$\frac{1}{\pi}\int_{-\pi}^{\pi} F_m(\theta)\,d\theta = 1, \quad \frac{1}{\pi}\int_{-\pi}^{\pi} P(r,\theta)\,d\theta = 1.$$

5.4. *K*-methods and their kernels. In what follows we shall suppose throughout that T is linear and regular, but we shall find it convenient to restrict the method further. We suppose that

(5.4.1) $$\sum_n n\,|\alpha_{m,n}| < \infty$$

for every m. We call such a method a *K*-method. We denote the transform of $D_n(\theta)$ by

(5.4.2) $$K_m(\theta) = \sum_n \alpha_{m,n}\frac{\sin(n+\frac{1}{2})\theta}{2\sin\frac{1}{2}\theta} = \sum_n \alpha_{m,n} D_n(\theta),$$

which we call the *kernel* of T; and the transform of the partial sum of the F.s. of $f(t)$, for $t=\theta$, by

(5.4.3)
$$\tau_m(\theta) = \tau_m(\theta,f) = \sum_n \alpha_{m,n}s_n(\theta,f) = \frac{1}{\pi}\sum_n \alpha_{m,n}\int_{-\pi}^{\pi} f(\theta+t)\,D_n(t)\,dt.$$

* We used C and S differently in §§ 3.10–11.

Then

$$(5.4.4)\quad \tau_m(\theta) = \frac{1}{\pi}\int_{-\pi}^{\pi} f(\theta+t)\,K_m(t)\,dt = \frac{2}{\pi}\int_0^{\pi} \phi(\theta,t)\,K_m(t)\,dt,$$

the term-by-term integration being justified by the convergence of

$$\int_{-\pi}^{\pi} |\,f(\theta+t)\,|\,\{\sum_n |\,\alpha_{m,n}\,|\,|\,D_n(t)\,|\}\,dt \leqq \sum_n (n+\tfrac{1}{2})\,|\,\alpha_{m,n}\,|\int_{-\pi}^{\pi} |\,f(t)\,|\,dt.$$

From (5.4.4) we can deduce tests for the summability of the F.s. analogous to those for convergence in §§ 4.2–4.3. If $f = 1$, then $s_n = 1$ for all n, and $\sigma_m = \Sigma\alpha_{m,n} = r_m \to 1$. Hence

$$(5.4.5)\qquad \frac{2}{\pi}\int_0^{\pi} K_m(t)\,dt = r_m \to 1.$$

Also

$$(5.4.6)\qquad \int_\delta^{\pi} \phi D_n\,dt \to 0,$$

by Theorem 31, for any positive δ and uniformly in θ; and therefore, since the method is regular,

$$(5.4.7)\qquad \int_\delta^{\pi} \phi K_m\,dt = \sum_n \alpha_{m,n}\int_\delta^{\pi} \phi D_n\,dt \to 0,$$

uniformly in θ.* In particular

$$(5.4.8)\qquad \int_\delta^{\pi} K_m\,dt \to 0.$$

Combining (5.4.5)–(5.4.8) with (5.4.4), we obtain

$$(5.4.9)\qquad \tau_m(\theta) - c = \frac{2}{\pi}\int_0^{\delta} g_c(t)\,K_m(t)\,dt + o(1),$$

where $g_c(t)$ is defined as in (4.2.3). Finally we observe that this is true uniformly, for any fixed positive δ, in any set of θ in which $c = c(\theta)$ is bounded.

Theorem 69. *In order that the Fourier series of $f(t)$, for $t = \theta$, should be summable* (T) *to c, it is necessary and sufficient that*

$$(5.4.10)\qquad J(\delta, c, m) = \int_0^{\delta} g_c(t)\,K_m(t)\,dt \to 0$$

for a positive δ. If (5.4.10) *holds uniformly in a set E in which $c = c(\theta)$ is bounded, then the convergence is uniform in E.*

There is naturally a variant of Theorem 69 with a continuous parameter.

* So that the conclusion still holds if $\phi(\theta, t)$ is replaced by $\phi(\theta + h_m, t)$ and $h_m \to 0$.

5.5. Summability of the Fourier series at a point of continuity or jump. The first clause of our next theorem is one of the fundamental theorems of the subject.

Theorem 70. *Suppose that* T *is a K-method and that*

$$\frac{2}{\pi}\int_0^{\pi} |K_m(t)|\,dt \leqq H, \tag{5.5.1}$$

where H is independent of m. Then

(i) *if* $\phi(+0) = \lim\limits_{t\to 0} \tfrac{1}{2}\{f(\theta+t)+f(\theta-t)\}$ *exists, the series is summable to* $\phi(+0)$; *in particular it is summable to* $f(\theta)$ *at a point of continuity and to* $\tfrac{1}{2}\{f(\theta+0)+f(\theta-0)\}$ *at a point of jump*:

(ii) *more generally, if* $h_m \to 0$, *then*

$$\tau_m(\theta+h_m) \to f(\theta) \tag{5.5.2}$$

at a point of continuity, and

(5.5.3)

$$\tau_m(\theta+h_m) - \frac{1}{\pi}\{f(\theta+0)-f(\theta-0)\}\int_0^{h_m} K_m(t)\,dt \to \tfrac{1}{2}\{f(\theta+0)+f(\theta-0)\}$$

at a point of jump:

(iii) *if f is continuous in a closed interval* $\langle a, b\rangle$, *then the series is uniformly summable* (T), *to f, in* $\langle a, b\rangle$; *and if f is bounded in* $\langle a, b\rangle$, *then* $\tau_m(\theta)$ *is bounded in any interior interval*:

(iv) $$|\tau_m(\theta)| \leqq H \operatorname{Max} |f(\theta)|, \quad \frac{1}{\pi}\int_{-\pi}^{\pi} |\tau_m(\theta)|\,d\theta \leqq \frac{H}{\pi}\int_{-\pi}^{\pi} |f(\theta)|\,d\theta,$$

where H has the same value as in (5.5.1): *and*

(v) $$\tau_m(\theta) \to f(\theta) \quad (L).$$

It follows from (5.4.5) that the condition (5.5.1) is satisfied whenever $K_m(t) \geqq 0$ for all m and t. This is true, in particular, after (5.3.5), for the $(C, 1)$ and (A) methods. After (5.4.5), $H \geqq 1$.

(i) If $\phi(+0)$ exists, and $c = \phi(+0)$, we can choose $\eta < \delta$ so that $|g_c(t)| < \epsilon$ for $|t| \leqq \eta$. Then

$$|J(\delta, c, m)| \leqq \epsilon\int_0^{\eta} |K_m(t)|\,dt + \left|\int_{\eta}^{\delta} g_c(t)\,K_m(t)\,dt\right| \leqq \tfrac{1}{2}\pi H\epsilon + o(1),$$

by (5.5.1) and (5.4.7); and (5.4.10) follows.

(ii) Suppose first that f is continuous at θ and that $h_m \to 0$. Then

$$|\phi(\theta+h_m, t) - f(\theta)| = \tfrac{1}{2}\,|f(\theta+h_m+t)+f(\theta+h_m-t)-2f(\theta)| < \epsilon$$

if $|t| \leqq \delta(\epsilon)$ and $m \geqq M(\epsilon)$. After (5.4.4), $\tau_m(\theta+h_m)-f(\theta)$ is

$$\frac{2}{\pi}\int_0^\delta \{\phi-f(\theta)\}K_m\,dt+\frac{2}{\pi}\int_\delta^\pi \phi K_m\,dt-\frac{2f(\theta)}{\pi}\int_\delta^\pi K_m\,dt+o(1),$$

and the second and third terms tend to 0, by (5.4.7) and (5.4.8). Hence

$$|\tau_m(\theta+h_m)-f(\theta)| \leqq \frac{2\epsilon}{\pi}\int_0^\delta |K_m(t)|\,dt+o(1) \leqq H\epsilon+o(1);$$

and therefore $\qquad \tau_m(\theta+h_m)\to f(\theta).$

If θ is a point of jump, we may suppose that

$$f(\theta) = \tfrac{1}{2}\{f(\theta+0)+f(\theta-0)\}$$

and $h_m > 0$. We write $d = f(\theta+0)-f(\theta-0)$ and

$$g(t) = f(t)-\frac{d}{\pi}\mathbf{f}(t-\theta) = f(t)-k(t)$$

(eliminating the discontinuity as in § 3.12). Then g is continuous at θ, and so

$$(5.5.4) \qquad \tau_m(\theta+h_m, g)\to g(\theta) = \tfrac{1}{2}\{f(\theta+0)+f(\theta-0)\}.$$

Also $\qquad \tau_m(\theta+h_m, k) = \dfrac{2}{\pi}\displaystyle\int_0^\pi \chi K_m\,dt,$

where

$$\chi = \tfrac{1}{2}\{k(\theta+h_m+t)+k(\theta+h_m-t)\} = \frac{d}{2\pi}\{\mathbf{f}(h_m+t)+\mathbf{f}(h_m-t)\},$$

by (5.4.4), so that

$$\chi = \frac{d}{2\pi}(\pi-h_m) \quad (0<t<h_m), \qquad \chi = -\frac{d}{2\pi}h_m \quad (h_m<t<\pi)$$

if h_m is small. Hence

$$(5.5.5) \quad \tau_m(\theta+h_m, k) = \frac{d(\pi-h_m)}{\pi^2}\int_0^{h_m} K_m\,dt-\frac{dh_m}{\pi^2}\int_{h_m}^\pi K_m\,dt$$
$$= \frac{d}{\pi}\int_0^{h_m} K_m\,dt+o(1);$$

and (5.5.3) follows from (5.5.4) and (5.5.5).

(iii) If f is continuous (and so uniformly continuous) in $\langle a,b\rangle$, we can apply the argument of (i), uniformly, to $\tau_m(\theta)-f(\theta)$. The proof of boundedness is similar but simpler.

(iv) It follows from (5.4.4) that

$$|\tau_m(\theta)| \leqq \frac{1}{\pi}\int_{-\pi}^\pi |f(\theta+t)|\,|K_m(t)|\,dt,$$

and the two inequalities (the first of which is significant only if

f is L^∞) are corollaries. Finally, we can choose a continuous $f^*(\theta)$ so that

$$(5.5.6)\quad \int_{-\pi}^{\pi} |f(\theta+t)-f^*(\theta+t)|\,d\theta = \int_{-\pi}^{\pi} |f(\theta)-f^*(\theta)|\,d\theta \leqq \frac{\epsilon}{H} \leqq \epsilon,$$

for every t. Then, after (5.4.4), (5.5.6), and (5.5.1),

$$|\tau_m(\theta,f)-\tau_m(\theta,f^*)| \leqq \frac{1}{\pi}\int_{-\pi}^{\pi} |f(\theta+t)-f^*(\theta+t)|\,|K_m(t)|\,dt,$$

$$(5.5.7)\quad \int_{-\pi}^{\pi} |\tau_m(\theta,f)-\tau_m(\theta,f^*)|\,d\theta$$

$$\leqq \frac{1}{\pi}\int_{-\pi}^{\pi} |K_m(t)|\,dt \int_{-\pi}^{\pi} |f(\theta+t)-f^*(\theta+t)|\,d\theta \leqq \epsilon.$$

From (5.5.6) and (5.5.7) it follows that

$$\int_{-\pi}^{\pi} |\tau_m(\theta,f)-f(\theta)|\,d\theta \leqq \int_{-\pi}^{\pi} |\tau_m(\theta,f^*)-f^*(\theta)|\,d\theta + 2\epsilon.$$

But $\tau_m(\theta,f^*)\to f^*(\theta)$ uniformly, by (iii); and therefore

$$\int_{-\pi}^{\pi} |\tau_m(\theta,f)-f(\theta)|\,d\theta \leqq 3\epsilon$$

for $m \geqq M(\epsilon)$.

The integrals in (5.5.1) are the 'Lebesgue constants of the method T'. Their boundedness is also *necessary* for the validity of Theorem 70: if they are unbounded we can show, by a construction like that of § 4.13, that there are continuous functions whose F.s. are not always summable (T).

We have already observed that (5.5.1) is satisfied when $K_m(t) \geqq 0$. In this case the argument used in proving (i) shows that

$$\varliminf_{t\to 0} \phi(\theta,t) \leqq \varliminf_{m\to\infty} \tau_m(\theta) \leqq \varlimsup_{m\to\infty} \tau_m(\theta) \leqq \varlimsup_{t\to 0} \phi(\theta,t).$$

If also $r_m = 1$, and $\mu \leqq f(\theta) \leqq M$ for all θ, then $\mu \leqq \tau_m(\theta) \leqq M$. These conditions (or their variants) are satisfied by the $(C, 1)$ and A methods.

5.6. Summability almost everywhere. The condition (5.5.1) does not enable us to prove that the method is 'Fourier-effective' in the sense of § 5.1. In our next two theorems we impose stronger conditions on $K_m(t)$.

Theorem 71. *Suppose that* T *is a K-method, and that*

$$(5.6.1) \qquad \frac{2}{\pi}\int_0^{\pi} t\,|\,K_m'(t)\,|\,dt \leqq H.$$

Then $\tau_m(\theta)\to c$ *for any* θ *for which*

$$(5.6.2) \qquad \Phi(t) = \int_0^t g_c(u)\,du = o(t)$$

(*i.e. whenever* f *satisfies* l_c). *In particular*

$$(5.6.3) \qquad \tau_m(\theta)\to f(\theta)$$

for almost all θ.

It follows from (5.4.1) and (5.4.2) that $K_m(t)$ has a derivative continuous except perhaps at the origin; and it is easily verified that (5.6.1) implies (5.5.1), with $R+H$ for H.

Supposing now that the conditions of the theorem are satisfied, we can choose δ so that $|\,\Phi(t)\,|\leqq \epsilon t$ for $0<t\leqq\delta$. Then

$$J(\delta,c,m) = \int_0^{\delta} g_c(t)\,K_m(t)\,dt = \Phi(\delta)\,K_m(\delta) - \int_0^{\delta}\Phi(t)\,K_m'(t)\,dt;$$

and $|\,K_m(\delta)\,|\leqq \frac{1}{2}R\operatorname{cosec}\frac{1}{2}\delta \leqq \frac{1}{2}\pi R/\delta$, by (5.4.2) and (3.6.6). Hence

$$|\,J\,|\leqq \tfrac{1}{2}\pi R\epsilon + \epsilon\int_0^{\delta} t\,|\,K_m'\,|\,dt \leqq \tfrac{1}{2}\pi(R+H)\,\epsilon,$$

and the conclusion follows.

We shall see later that the A method satisfies (5.6.1), whereas the $(C,1)$ method does not. Thus Theorem 71 will not prove that the $(C,1)$ method is 'Fourier-effective'. For this we need another theorem.

Theorem 72. *Suppose that* T *is a K-method, and that*

$$(5.6.4) \qquad |\,K_m(t)\,| \leqq K_m^*(t),$$

where K_m^* *is absolutely continuous, except perhaps at the origin, and*

$$(5.6.5) \qquad \int_0^{\pi} t\,|\,K_m^{*\prime}(t)\,|\,dt \leqq H_1.$$

Then $\tau_m(\theta)\to c$ *whenever*

$$(5.6.6) \qquad \Phi^*(t) = \int_0^t |\,g_c(u)\,|\,du = o(t)$$

(*i.e. whenever* f *satisfies* L_c). *In particular* $\tau_m(\theta)\to f(\theta)$ *for almost all* θ.

We note, first, that

$$K_m^*(t) = K_m^*(\pi) - \int_t^{\pi}\frac{\tau K_m^{*\prime}(\tau)}{\tau}\,d\tau \leqslant K_m^*(\pi) + H_1 t^{-1} \leqslant H_2 t^{-1}.$$

Hence, if we choose δ so that $|\Phi^*(t)| \leqq \epsilon t$ for $0<t\leqq\delta$, then

$$|J(\delta, c, m)| \leqq \int_0^\delta |g_c(t)|\,|K_m(t)|\,dt \leqq \int_0^\delta |g_c(t)|\,K_m^*(t)\,dt$$

$$= \Phi^*(\delta)\,K_m^*(\delta) - \int_0^\delta \Phi^*(t)\,K_m^{*\prime}(t)\,dt \leqq \epsilon H_2 + \epsilon\int_0^\delta t\,|K_m^{*\prime}(t)|\,dt \leqq H\epsilon,$$

where $H = H_1 + H_2$; and this proves the theorem.

5.7. The $(C, 1)$ **summability of the Fourier series.** The kernel of the $(C, 1)$ method is $F_m(t)$, by (5.3.1) and (5.4.2); and it satisfies (5.5.1), by (5.3.6). Further, it is $O(m)$ in $(0, \pi/m)$ and $O(m^{-1}t^{-2})$ in $(\pi/m, \pi)$. This gives an alternative proof that it satisfies (5.5.1), and also shows that

$$|F_m| \leqq F_m^* = \frac{Hm}{1+m^2t^2}, \tag{5.7.1}$$

with an appropriate H. Since

$$\int_0^\pi t\,|F_m^{*\prime}|\,dt = 2H\int_0^\pi \frac{m^3t^2\,dt}{(1+m^2t^2)^2} < 2H\int_0^\infty \frac{u^2\,du}{(1+u^2)^2}, \tag{5.7.2}$$

it follows that F_m also satisfies the conditions of Theorem 72.

Theorem 73. *The* $(C,1)$ *method satisfies the conditions of Theorems* 70 *and* 72. *In particular, the Fourier series of* $f(t)$, *for* $t=\theta$, *is summable* $(C,1)$ *to* $f(\theta)$ *at a point of continuity, and to* $\frac{1}{2}\{f(\theta+0)+f(\theta-0)\}$ *at a point of jump, and is uniformly summable in any closed interval of continuity. More generally, it is summable to c at any point at which* $f(t)$ *satisfies* L_c, *and to* $f(\theta)$ *for almost all* θ. *Also*

$$\sigma_m(\theta) \to f(\theta) \quad (L). \tag{5.7.3}$$

It is important to observe that $F_m(t)$ does *not* satisfy the conditions of Theorem 71. In fact

$$F_m'(t) = \frac{\sin(m+1)t}{4\sin^2\frac{1}{2}t} - \frac{\sin^2\frac{1}{2}(m+1)t\cos\frac{1}{2}t}{2(m+1)\sin^3\frac{1}{2}t}.$$

The contribution of the second term to $\int t\,|K_m'|\,dt$ is bounded, but that of the first behaves like a multiple of the Lebesgue constant of § 4.14.

In applying clause (ii) of Theorem 70, it is convenient to suppose that $mh_m \to a$, where $-\infty \leqq a \leqq \infty$. Then, writing M for $m+1$, we have

$$\int_0^{h_m} F_m(t)\,dt = \frac{1}{2M}\int_0^{h_m}\left(\frac{\sin\frac{1}{2}Mt}{\sin\frac{1}{2}t}\right)^2 dt = \frac{2}{M}\int_0^{h_m}\frac{\sin^2\frac{1}{2}Mt}{t^2}\,dt + o(1)$$

$$= \int_0^{\frac{1}{2}Mh_m}\left(\frac{\sin u}{u}\right)^2 du + o(1) = \int_0^{\frac{1}{2}a}\left(\frac{\sin u}{u}\right)^2 du + o(1).$$

Hence (5.5.3) takes the form

(5.7.4)

$$\sigma_m(\theta+h_m)\to\tfrac{1}{2}\{f(\theta+0)+f(\theta-0)\}+\frac{1}{\pi}\{f(\theta+0)-f(\theta-0)\}\int_0^{\frac{1}{2}a}\left(\frac{\sin u}{u}\right)^2du.$$

In particular

$$\sigma_m(\theta+h_m)\to f(\theta+0) \tag{5.7.5}$$

when $a=\infty$ (the integral having then the value $\frac{1}{2}\pi$).

The equation (5.7.4) has an interesting application to the theory of the 'Gibbs phenomenon' (§ 3.12). We may define the 'Gibbs set corresponding to T', at a point θ of jump of $f(t)$, as we defined the ordinary Gibbs set in § 3.12, viz. as the aggregate of all limits of $\tau_m(\xi)$ when $m\to\infty$ and $\xi\to\theta$. This is the same as the aggregate of all the limits of $\tau_m(\theta+h_m)$ corresponding to all values of a, and (5.7.4) shows that in this case these limits just fill up the interval from $f(\theta-0)$ to $f(\theta+0)$. Thus *the* $(C, 1)$ *method has no Gibbs phenomenon*. This is a general feature of methods with a positive kernel.

5.8. The $(C, 1)$ summability of the conjugate series. There is a general theory of summability of the c.s. analogous to that developed for the F.s. in §§ 5.4–5.6. This we must leave to the reader; but the main theorem about $(C, 1)$ summability is important and will be wanted later. We write

$$\tilde{\sigma}_m(\theta)=\tilde{\sigma}_m(\theta,f)=\frac{1}{m+1}\sum_0^m\tilde{s}_n(\theta,f). \tag{5.8.1}$$

Theorem 74. *If f satisfies $\tilde{L}_0$ for $t=\theta$, then*

$$\tilde{\sigma}_m(\theta)-\tilde{f}_m(\theta)=\tilde{\sigma}_m(\theta)-\frac{1}{2\pi}\int_{\pi/m}^{\pi}\psi(\theta,t)\cot\tfrac{1}{2}t\,dt\to 0, \tag{5.8.2}$$

and this is true for almost all θ. If also $\tilde{f}(\theta)$ exists, then

$$\tilde{\sigma}_m(\theta)\to\tilde{f}(\theta). \tag{5.8.3}$$

The last two clauses are corollaries of the first. We shall prove in § 6.8 (Theorem 89) that $\tilde{f}(\theta)$ exists p.p., so that (5.8.3) is true p.p.; but Theorem 89 lies deeper. We write

$$\Psi^*(t)=\int_0^t|\psi(u)|\,du. \tag{5.8.4}$$

Since f satisfies $\tilde{L}_0$, $\Psi^*(t)=o(t)$. After (4.9.1), (5.3.3) and (5.8.1), we have

$$\begin{aligned}\tilde{\sigma}_m(\theta)&=\frac{1}{m+1}\sum_0^m\frac{1}{\pi}\int_0^{\pi}\psi(\theta,t)\tilde{D}_m(t)\,dt=\frac{1}{\pi}\int_0^{\pi}\psi(\theta,t)\,G_m(t)\,dt\\&=\frac{1}{\pi}\int_0^{\pi}\psi\left\{\tfrac{1}{2}\cot\tfrac{1}{2}t-\frac{\sin(m+1)t}{4(m+1)\sin^2\frac{1}{2}t}\right\}dt,\end{aligned}$$

$$\tilde{\sigma}_m - \tilde{f}_m = \frac{1}{\pi}\int_0^{\pi/m} \psi G_m \, dt - \frac{1}{4(m+1)\pi}\int_{\pi/m}^{\pi} \psi \frac{\sin(m+1)t}{\sin^2 \frac{1}{2}t} \, dt = J_1 - J_2,$$

say. Now

$$G_m = \frac{1}{(2\sin\frac{1}{2}t)^2}\left\{\sin t - \frac{\sin(m+1)t}{m+1}\right\} = O(m^2 t)$$

for $0 < t \leqq \pi/m$; and so

$$J_1 = O\left(m^2 \int_0^{\pi/m} t\,|\psi|\,dt\right) = O\left\{m\Psi^*\left(\frac{\pi}{m}\right)\right\} = o(1).$$

On the other hand

$$|J_2| \leqq \frac{\pi}{4(m+1)}\int_{\pi/m}^{\pi} |\psi| \frac{dt}{t^2} \leqq \frac{\pi}{4(m+1)}\left\{\frac{\Psi^*(\pi)}{\pi^2} + 2\int_{\pi/m}^{\pi} \frac{\Psi^*(t)}{t^3}\,dt\right\}$$

$$= o(1) + O\left(\frac{1}{m}\right)\int_{\pi/m}^{\pi} o\left(\frac{1}{t^2}\right) dt = o(1);$$

so that $\tilde{\sigma}_m - \tilde{f}_m = o(1)$.

5.9. Summability (A). If Σu_n is summable $(C, 1)$, it is certainly summable (A). It is however important to consider A summability independently, since (a) it succeeds under wider conditions and (b) it helps to link up the theory of F.s. with the theory of harmonic and analytic functions.

The kernel of the A method is the $P(r, \theta)$ of (5.3.2). If f is L, then

(5.9.1)

$$u(r,\theta) = \tfrac{1}{2}a_0 + \sum_1^{\infty} A_n(\theta)\, r^n = \frac{1}{\pi}\int_{-\pi}^{\pi} \{\tfrac{1}{2} + r\cos(t-\theta) + \ldots\} f(t)\, dt$$

$$= \frac{1}{2\pi}\int_{-\pi}^{\pi} \frac{1-r^2}{1-2r\cos(t-\theta)+r^2} f(t)\, dt = \frac{1}{\pi}\int_{-\pi}^{\pi} P(r, t-\theta) f(t)\, dt.$$

Similarly, if $Q(r, \theta)$ is defined by (5.3.4),

$$(5.9.2) \qquad v(r,\theta) = \sum_1^{\infty} B_n(\theta)\, r^n = \frac{1}{\pi}\int_{-\pi}^{\pi} Q(r, t-\theta) f(t)\, dt.$$

These two functions are conjugate harmonic functions in the sense of § 1.2. The function $u(r, \theta)$ associated with f is often called the *Poisson integral* of f. The theory of the A summability of F.s. or c.s. is the theory of radial limits of harmonic functions generated in this particular way.

If we start from an *arbitrary* harmonic function regular for $r < 1$, then it may be possible to define a *boundary function* associated with it (e.g. as a radial limit). This boundary function may or may not be L. If it is L, and we call it f, we can define the F.s. and Poisson integral of f; and the question then arises whether this Poisson integral is the harmonic function from which we started. That the answer is not necessarily affirmative is shown by the example of the function $P(r, \theta)$, whose boundary function is 0 for $0 < \theta < 2\pi$. Questions of this kind are often difficult, and belong properly to the 'uniqueness theory' of which we give an introductory account in Ch. VII.

The most important difference between the $(C, 1)$ and A kernels is that the latter satisfies the conditions of Theorem 71. For $P \geqq 0$ and $P' \leqq 0$, and so

$$\int_0^\pi t \mid P' \mid dt = -\int_0^\pi tP'\, dt = \int_0^\pi P\, dt - \pi P(\pi) \leqq \int_0^\pi P\, dt = \tfrac{1}{2}\pi.$$

Theorem 75. *The A method satisfies the conditions of Theorems* 70 *and* 71. *In particular, the Fourier series of $f(t)$, for $t = \theta$, is summable (A) to $f(\theta)$ at a point of continuity and to $\frac{1}{2}\{f(\theta+0)+f(\theta-0)\}$ at a point of jump, and is uniformly summable in any closed interval of continuity. More generally, it is summable to c at any point at which $f(t)$ satisfies l_c, and to $f(\theta)$ almost everywhere. Also*

$$u(r, \theta) \to f(\theta) \quad (L). \tag{5.9.3}$$

If $h_r \to 0$ and $(1-r)^{-1} h_r \to a$, then

$$\frac{1}{2}\int_0^{h_r} \frac{1-r^2}{1-2r\cos t+r^2}\, dt = \arctan\left(\frac{1+r}{1-r}\tan \tfrac{1}{2}h_r\right) \to \arctan a.$$

Thus (5.5.3), with the modifications appropriate to the continuous parameter, gives

$$u(r, \theta+h_r) \to \tfrac{1}{2}\{f(\theta+0)+f(\theta-0)\} + \frac{1}{\pi}\{f(\theta+0)-f(\theta-0)\}\arctan a, \tag{5.9.4}$$

when $h_r \to 0$, $(1-r)^{-1} h_r \to a$. This becomes

$$u(r, \theta+h_r) \to f(\theta+0), \tag{5.9.5}$$

when $a = \infty$.

This result, which corresponds to (5.7.5), is proved, like (5.7.5), for a point of jump. It is natural to ask when (5.9.5) can be deduced from the existence of $f(\theta+0)$ only, and we shall prove that (5.9.5) *is true whenever $h_r \to 0$, $\underline{\lim}\,(1-r)^{-\frac{1}{2}} h_r = L > 0$ and $f(\theta+0)$ exists.*

We may suppose that $\theta = 0$ and $f(\theta+0) = f(+0) = 0$. It will plainly be enough to prove that

$$|J| = \left|\int_{-\delta}^{\delta} f(t)\, P(r, t-h_r)\, dt\right| < \epsilon \tag{5.9.6}$$

for a $\delta = \delta(\epsilon)$ and $1-r \leqq \rho(\epsilon)$. We can choose η and δ so that

$$\left(\frac{2}{L^2}+\pi\right)\eta<\epsilon, \quad |f(t)|<\eta \ \ (0<t\leqq\delta), \quad \int_{-\delta}^{0}|f|\,dt<\eta.$$

Then
$$P(r,t-h_r) \leqq \frac{1-r^2}{4r\sin^2\frac{1}{2}(t-h_r)} \leqq \frac{1-r}{2r\sin^2\frac{1}{2}h_r}$$

in $(-\delta, 0)$, and so

$$|J| \leqq \frac{1-r}{2r\sin^2\frac{1}{2}h_r}\int_{-\delta}^{0}|f|\,dt+\eta\int_{0}^{\delta}P(r,t-h_r)\,dt<\left(\frac{1-r}{2r\sin^2\frac{1}{2}h_r}+\pi\right)\eta,$$

$$\overline{\lim_{r\to 1}}\,|J| \leqq \left(\frac{2}{L^2}+\pi\right)\eta<\epsilon,$$

which is equivalent to (5.9.6).

It is more illuminating to state (5.9.4) or (5.9.5) as a theorem about harmonic functions. Suppose that P and P_0 are the points whose polar coordinates are (r,θ) and $(1,\theta_0)$, and that $\rho = 1-r$ and $h = \theta-\theta_0$. If $r\to 1$ and $\theta\to\theta_0$ in such a way that $h\sim a\rho$, then P tends to P_0 along a path making an angle $\operatorname{arc\,tan} a$ with the radius vector OP_0. If $\overline{\lim}\,\rho^{-\frac{1}{2}}h>0$, then P tends to P_0 in a cuspidal region bounded by the unit circle and a smaller circle touching it at P_0. In the first case $u(r,\theta)$ tends to the limit (5.9.4), in the second to $f(\theta_0+0)$, provided only that these limits exist.

5.10. The A summability of the conjugate series.

The theorem for the c.s. corresponding to Theorem 75 is

Theorem 76. *If f satisfies $\tilde{l}_0$, for $t = \theta$, and* $\eta = \operatorname{arc\,sin}(1-r)$, *then*

$$(5.10.1)\quad v(r,\theta)-\tilde{f}_\eta(\theta) = v(r,\theta)-\frac{1}{2\pi}\int_{\eta}^{\pi}\psi(\theta,t)\cot\tfrac{1}{2}t\,dt\to 0.$$

If $\tilde{f}(\theta)$ exists, then $v(r,\theta)\to\tilde{f}(\theta)$.

It follows from (5.9.2) that

$$v(r,\theta) = \frac{1}{\pi}\int_{-\pi}^{\pi}f(\theta+t)\,Q(r,t)\,dt = \frac{1}{\pi}\int_{0}^{\pi}\psi(\theta,t)\,Q(r,t)\,dt,$$

$$(5.10.2)\quad v(r,\theta)-\tilde{f}_\eta(\theta) = \frac{1}{\pi}\int_{0}^{\eta}\psi Q\,dt-\frac{1}{\pi}\int_{\eta}^{\pi}\psi q\,dt = J_1-J_2,$$

say, where

$$Q(t) = \frac{r\sin t}{1-2r\cos t+r^2}, \quad q(t) = \tfrac{1}{2}\cot\tfrac{1}{2}t-Q(t) = \frac{\frac{1}{2}(1-r)^2\cot\frac{1}{2}t}{1-2r\cos t+r^2}.$$

We write $\Delta = 1-2r\cos t+r^2 = (1-r)^2+4r\sin^2\frac{1}{2}t.$

Then it is easily verified that

$$Q(\pi) = 0, \quad q(\pi) = 0, \quad Q(\eta) = O\left(\frac{1}{\eta}\right), \quad q(\eta) = O\left(\frac{1}{\eta}\right),$$

$$Q' = \frac{r\{(1+r^2)\cos t - 2r\}}{\Delta^2} = O\left(\frac{1}{\eta^2}\right) \quad (0<t \leqq \eta),$$

$$q' = -\frac{(1-r)^2 \operatorname{cosec}^2 \frac{1}{2}t}{4\Delta} - \frac{2r(1-r)^2 \cos^2 \frac{1}{2}t}{\Delta^2} = O\left\{\frac{(1-r)^2}{t^4}\right\} = O\left(\frac{\eta^2}{t^4}\right).$$

Hence, if $\Psi(t) = \int_0^t \psi\, du$, so that $\Psi(t) = o(t)$, we have

$$J_1 = \frac{1}{\pi}\Psi(\eta)\, Q(\eta) - \frac{1}{\pi}\int_0^\eta \Psi Q'\, dt = o(1) + \frac{1}{\eta^2}\int_0^\eta o(t)\, dt = o(1),$$

$$J_2 = -\frac{1}{\pi}\Psi(\eta)\, q(\eta) - \frac{1}{\pi}\int_\eta^\pi \Psi q'\, dt = o(1) + \eta^2 \int_\eta^\pi \frac{o(t)}{t^4}\, dt = o(1),$$

and the theorem follows from (5.10.2).

There is one important difference between Theorems 74 and 76. If $\tilde{f}(\theta)$ exists, then (after the last paragraph of § 4.10) f certainly satisfies $\tilde{l}_0$: thus the existence of $\tilde{f}(\theta)$ is sufficient in itself to ensure the summability (A) of the c.s. to $\tilde{f}(\theta)$. In Theorem 74 we required the stronger condition $\tilde{L}_0$; this is of course satisfied if $\tilde{f}(\theta)$ exists *as a Lebesgue integral.*

5.11. Some applications of Theorems 70—76. Our theorems concerning summability enable us to make a number of important additions to the results of our earlier chapters, of which we select three.

(1) The first follows immediately from the regularity of the A method: if a series is convergent, it is summable (A) to the same sum. Hence we deduce

Theorem 77. *If the Fourier series of $f(t)$ converges at a point θ where $f(t)$ satisfies l_c, then its sum is c. In particular, if it converges at a point of continuity or jump, then its sum is $f(\theta)$ or $\frac{1}{2}\{f(\theta+0)+f(\theta-0)\}$. If it converges almost everywhere, then it represents $f(\theta)$ almost everywhere.*

If the conjugate series converges at a point where $\tilde{f}(\theta)$ exists, then its sum is $\tilde{f}(\theta)$.

When we have proved (Theorem 89) that $\tilde{f}$ exists p.p., it will follow that, if the c.s. converges p.p., it represents $\tilde{f}$ p.p.

(2) The second concerns functions whose F.c. are $O(n^{-1})$: by Theorem 37, all functions of V satisfy this condition. The application depends on the 'Tauberian' theorem that *if Σu_n is summable*

$(C, 1)$, *and* $u_n = O(n^{-1})$, *then* Σu_n *is convergent.* It follows that the F.s. of such a function is convergent whenever it is summable $(C, 1)$, and converges to $f(\theta)$ p.p.

In particular, when f is V, the Fourier series of f converges to $\frac{1}{2}\{f(\theta+0)+f(\theta-0)\}$. This result, combined with Theorem 52, leads to alternative proofs of Theorems 56 and 57.

(3) It is easy to deduce Weierstrass's theorem (Theorem 23) and the completeness theorem (Theorem 19) from Theorem 73.

(i) If f is continuous and has period 2π, then $\sigma_m(\theta)$ tends uniformly to $f(\theta)$. This proves Theorem 24, and Theorem 23 is a corollary, as we saw in § 2.9.

(ii) In proving Theorem 19 we may (as we saw in § 2.6) assume that f is continuous. If all the F.c. of f are 0, then $\sigma_m(\theta) = 0$ for all θ, and therefore its limit is 0. Alternatively (dispensing with the preliminary reduction to the continuous case) $\sigma_m(\theta) \to f(\theta)$ p.p., and therefore $f(\theta) \equiv 0$. This proof is a little less elementary because it depends on Theorem 72 instead of Theorem 70.

5.12. Derived series of Fourier series. If $f_1(t)$ is an integral of $f(t)$, then

$$\frac{\Phi(t)}{t} = \frac{1}{t}\int_0^t g_c(u)\,du$$

$$= \frac{1}{2t}\int_0^t \{f(\theta+u)+f(\theta-u)-2c\}\,du = \frac{f_1(\theta+t)-f_1(\theta-t)}{2t} - c.$$

Hence $\Phi(t) = o(t)$ is equivalent to

$$\frac{f_1(\theta+t)-f_1(\theta-t)}{2t} \to c.$$

In these circumstances we say that $f_1(t)$ has a *generalized first derivative*, for $t = \theta$, equal to c, and write $Df_1(\theta) = c$. It follows from Theorem 75 that *the F.s. of* $f(t)$ *is summable* (A) *for any* θ *for which its integral* $f_1(t)$ *has a generalized first derivative.*

The F.s. of f is the derived series of the F.s. of f_1, i.e. the series obtained from it by formal differentiation; and so the result just proved suggests the more general theorem which follows, in which f takes the place of f_1 and the derived series is not usually itself a F.s.

Theorem 78. *If $f(t)$ has a generalized first derivative $Df(\theta)$ for $t=\theta$, then*

$$\frac{\partial u(r,\theta)}{\partial\theta}=\sum_1^\infty n(b_n\cos n\theta-a_n\sin n\theta)\,r^n\to Df(\theta), \tag{5.12.1}$$

when $r\to 1$. In other words, the derived series of the Fourier series of $f(t)$, for $t=\theta$, is summable (A) to sum $Df(\theta)$.

In fact, differentiating (5.9.1) with respect to θ,

$$\begin{aligned}\frac{\partial u}{\partial\theta}&=-\frac{1}{\pi}\int_{-\pi}^{\pi}f(t)\,P'(t-\theta)\,dt=-\frac{1}{\pi}\int_{-\pi}^{\pi}f(\theta+t)\,P'(t)\,dt\\&=\frac{1}{\pi}\int_0^{\pi}\frac{f(\theta+t)-f(\theta-t)}{2\sin t}\,L(t)\,dt,\end{aligned}$$

where $P(t)=P(r,t)$, dashes denote differentiations with respect to t, and

$$L(t)=-2\sin t\,P'(t)=\frac{2r(1-r^2)\sin^2 t}{(1-2r\cos t+r^2)^2}.$$

The first factor in the integrand tends to $Df(\theta)$ when $t\to 0$. Also $L(t)\geqq 0$, $L(t)\to 0$ uniformly in any interval $0<\delta\leqq t\leqq\pi$, and

$$\frac{1}{\pi}\int_0^{\pi}L(t)\,dt=\frac{2}{\pi}\int_0^{\pi}\cos t\,P(t)\,dt=r\to 1.$$

Hence (5.12.1) follows, by an argument substantially the same as that used in the proof of Theorem 70 (i).

The theorem is one of a scale. It can be shown, for example, that if f has k (ordinary) derivatives for $t=\theta$, then the kth partial derivative of u with respect to θ tends to $f^{(k)}(\theta)$; and these results can be interpreted as theorems concerning the summability (A) of the successive derived series of the F.s. of f.

VI. APPLICATIONS OF THE THEOREMS OF CHAPTER V

6.1. Introduction. In this chapter we collect a number of theorems which are mainly applications of those proved in Ch. v. Theorem 79 belongs logically to Ch. IV, but the proof depends on the properties of the function $F_m(\theta)$ of Ch. v, so that it is more convenient to insert it here.

6.2. A Fourier series which diverges almost everywhere. We proved in Ch. v (Theorems 73 and 75) that the $(C, 1)$ and A methods of summation are 'Fourier-effective', that is to say that they sum every F.s., to its generating function, p.p. We shall now prove that *classical convergence is not Fourier-effective.*

Theorem 79. *There are Fourier series which diverge almost everywhere.*

The kernel of the proof lies in the preliminary theorem which follows.

Theorem 80. *There is a sequence of trigonometrical polynomials ϕ_n with the properties*

$$\text{(i) } \phi_n \geqq 0, \qquad \text{(ii) } \frac{1}{\pi}\int_0^{2\pi} \phi_n \, d\theta = 1,$$

(iii) *to each ϕ_n correspond* (*a*) *an M_n tending to infinity,* (*b*) *a set E_n whose measure tends to 2π, and* (*c*) *an integer q_n such that*

$$|s_{p_n}(\theta, \phi_n)| > M_n$$

for every θ of E_n and a $p_n = p_n(\theta) \leqq q_n$.

We write $\qquad A_j = 4j\pi/(2n+1) \quad (0 \leqq j \leqq n);$

suppose that $m_0, m_1, \ldots, m_n$ are integers with the properties (*a*) $m_0 \geqq n^4$, (*b*) $m_{l+1} > 2m_l$ and (*c*) $2m_l + 1$ is divisible by $2n+1$; and define ϕ_n by

$$\phi_n(\theta) = \frac{1}{n+1}\{F_{m_0}(\theta - A_0) + F_{m_1}(\theta - A_1) + \ldots + F_{m_n}(\theta - A_n)\},$$

where F_m is Fejér's kernel (5.3.1). Plainly ϕ_n is a non-negative polynomial and, since every F_m begins with $\frac{1}{2}$, it satisfies (ii).

If $m_j \leqq k < m_{j+1}$, then

$$s_k(\theta, \phi_n) = \frac{1}{n+1}\left[\sum_{l=0}^{j} F_{m_l}(\theta - A_l) + \sum_{l=j+1}^{n}\left\{\tfrac{1}{2} + \sum_{r=1}^{k}\frac{m_l + 1 - r}{m_l + 1}\cos r(\theta - A_l)\right\}\right];$$

and if we separate the last term into two, by writing

$$m_l + 1 - r = k + 1 - r + (m_l - k),$$

we obtain

$$s_k(\theta, \phi_n) = \frac{1}{n+1}\sum_{l=0}^{j} F_{m_l}(\theta - A_l) + \frac{1}{n+1}\sum_{l=j+1}^{n}\frac{k+1}{m_l+1}F_k(\theta - A_l) \tag{6.2.1}$$

$$+ \frac{1}{n+1}\sum_{l=j+1}^{n}\frac{m_l - k}{m_l + 1}D_k(\theta - A_l) = S_1 + S_2 + S_3,$$

say. Here D_k is Dirichlet's kernel. The identity is true for $m_j \leqq k < m_{j+1}$: actually we take $k = m_j$.

We consider $s_{m_j}(\theta, \phi_n)$ in the interval $(A_j + n^{-2},\ A_{j+1} - n^{-2})$ or J_j. Since $F_m(t) = O(m^{-1}t^{-2})$ and $|\theta - A_l| \geqq n^{-2} \geqq m_l^{-\frac{1}{4}}$ for θ in J_j and all l, the F_m in (6.2.1) are uniformly bounded. Hence S_1 and S_2 are bounded and

$$(6.2.2) \qquad |s_{m_j}(\theta, \phi_n)| > |S_3| - H$$

for θ of J_j.

Now

$$(m_j + \tfrac{1}{2})(A_l - A_{j+1}) = (2m_j + 1)(l - j - 1)\frac{2\pi}{2n+1} \equiv 0 \pmod{2\pi}.$$

Hence $S_3 = \sigma T$, where

$$(6.2.3) \qquad \sigma = \sin\{(m_j + \tfrac{1}{2})(A_{j+1} - \theta)\},$$

$$(6.2.4) \qquad T = \frac{1}{n+1} \sum_{l=j+1}^{n} \frac{m_l - m_j}{m_l + 1} \frac{1}{2\sin\frac{1}{2}(A_l - \theta)}.$$

The terms of T are positive and, since $m_{j+1} > 2m_j$,

$$(6.2.5) \qquad (n+1)T > \frac{1}{2} \sum_{l=j+1}^{n} \frac{1}{A_l - \theta} > \frac{1}{2} \sum_{l=j+1}^{n} \frac{1}{A_l - A_j}$$

$$= \frac{2n+1}{8\pi} \sum_{l=j+1}^{n} \frac{1}{l-j} > \frac{2n+1}{8\pi} \log(n-j) > \frac{n+1}{16\pi} \log n,$$

if $j < n - \sqrt{n}$.

We define E_n as the set of θ for which θ lies in a J_j with $1 \leqq j < n - \sqrt{n}$ and

$$(6.2.6) \qquad |\sigma| = |\sin\{(m_j + \tfrac{1}{2})(A_{j+1} - \theta)\}| > (\log n)^{-\frac{1}{2}}.$$

Then, from (6.2.2)–(6.2.6),

$$|s_{m_j}(\theta, \phi)| > \frac{(\log n)^{\frac{1}{2}}}{16\pi} - H = M_n$$

for θ in E_n. We take this value for M_n, m_n for q_n, and m_j for $p_n(\theta)$ when θ is in J_j.

It remains only to prove that $mE_n \to 2\pi$; and this is trivial. For the J_j with $1 \leqq j \leqq n - n^{\frac{1}{2}}$ fill up all but $O(n \,.\, n^{-2}) + O(n^{\frac{1}{2}} \,.\, n^{-1}) = O(n^{-\frac{1}{2}})$ of $(0, 2\pi)$, and the part of them in which (6.2.6) is not satisfied is $O\{(\log n)^{-\frac{1}{2}}\}$ of the whole.

It is now easy to prove Theorem 79. Since $M_n \to \infty$, we can choose a sequence (n_s) so that $\Sigma M_{n_s}^{-\frac{1}{2}} < \infty$. We write ϕ_{n_s} as a sum of exponentials, and put

$$T_{n_s} = \frac{e^{r_s i\theta}}{M_{n_s}^{\frac{1}{2}}} \phi_{n_s} = \frac{e^{r_s i\theta}}{M_{n_s}^{\frac{1}{2}}} \sum_{-m_{n_s}}^{m_{n_s}} c_\nu e^{\nu i\theta},$$

$$\Phi = T_{n_1} + T_{n_2} + \ldots, \qquad T = (T_{n_1}) + (T_{n_2}) + \ldots,$$

(T_{n_s}) being T_{n_s} written out at length as a sum of exponentials. We can plainly choose the r_s so that there is no overlapping between the terms of T, when T is a (complex) t.s. Since

$$\frac{1}{\pi}\int_0^{2\pi} |e^{-pi\theta}|\,\Sigma\,|T_{n_s}|\,d\theta \leqq \frac{1}{\pi}\Sigma\frac{1}{M_{n_s}^{\frac{1}{2}}}\int_0^{2\pi}\phi_{n_s}\,d\theta = \Sigma\frac{1}{M_{n_s}^{\frac{1}{2}}} < \infty,$$

the series ΣT_{n_s} converges p.p. to an integrable Φ, and may be integrated term-by-term after multiplication by $e^{-pi\theta}$. Hence T is the F.s. of Φ. On the other hand, if θ is in E_{n_s}, (T_{n_s}) contains a block of successive terms whose sum is numerically greater than $M_{n_s}^{\frac{1}{2}}$. Hence T is divergent if θ lies in an infinity of E_{n_s}; and this is true p.p. because $mE_{n_s} \to 2\pi$.*

6.3. Fourier series with positive coefficients. The relations between the continuity of a function and the convergence of its F.s. are quite simple when the F.c. are positive. In §§ 3.10–3.11 we considered series with positive *and decreasing* coefficients; in this section we prove two theorems related to those proved there but of a more general character. It is convenient to consider even and odd functions separately.

Theorem 81. *If f is even and bounded, $f \sim (a_n, 0)$, and $a_n \geqq 0$, then $\Sigma a_n < \infty$ (so that f is continuous).*

For $\sigma_{2n}(0)$ is bounded, by Theorem 70 (iv); and

$$\tfrac{1}{2}s_n(0) \leqq \frac{n+1}{2n+1}s_n(0) \leqq \frac{1}{2n+1}\sum_{n}^{2n} s_m(0) \leqq \sigma_{2n}(0)$$

because $a_m \geqq 0$. Hence $\Sigma a_n < \infty$ (and the F.s. is uniformly convergent).

Actually we need only suppose f bounded near the origin.

Theorem 82. *If f is odd and bounded, $f \sim (0, b_n)$, and $b_n \geqq 0$, then $s_n(\theta)$ is uniformly bounded. If also f is continuous, then the Fourier series of f is uniformly convergent.*

(i) If $|f| \leqq 1$, then

$$|\sigma_N(\theta)| = \left|\sum_1^N \left(1-\frac{m}{N+1}\right) b_m \sin m\theta\right| \leqq 1, \tag{6.3.1}$$

* T is a complex series. We can obtain a real one, if we wish to, by taking the real part of T. But this requires a slight modification of the argument, and it is then necessary to use Theorem 92 of Ch. VII.

by Theorem 70 (iv). Taking $N = 2n$ and $\theta = \pi/4n$, when $\sin m\theta \geqq m/2n$ for $1 \leqq m \leqq 2n$, we obtain

$$0 \leqq \sum_{1}^{2n} \left(1 - \frac{m}{2n+1}\right) mb_m \leqq 2n.$$

Hence
$$0 \leqq \tfrac{1}{2} \sum_{1}^{n} mb_m \leqq 2n$$

and *a fortiori*

(6.3.2)
$$\left| \sum_{1}^{n} \frac{m}{n+1} b_m \sin m\theta \right| \leqq 4.$$

Finally, taking $N = n$ in (6.3.1), and combining it with (6.3.2), we obtain

$$\left| \sum_{1}^{n} b_m \sin m\theta \right| \leqq 5.$$

(ii) If f is continuous, then σ_n tends uniformly to f. We can therefore write $f = \sigma_n + g$, where $n = n(\epsilon)$ and $|g| < \epsilon$ for all θ. The coefficients of σ_n do not exceed the corresponding coefficients of f, and therefore those of g are non-negative. It follows from (i) that $|s_m(\theta, g)| < 5\epsilon$ for all m and θ. Also $s_p(\theta, \sigma_n) = s_q(\theta, \sigma_n) = \sigma_n(\theta)$ if $q > p \geqq n$; and then

$$|s_q(\theta, f) - s_p(\theta, f)| = |s_q(\theta, g) - s_p(\theta, g)| < 10\epsilon.$$

Hence the F.s. of f is uniformly convergent.

It is naturally not true in this case that f is necessarily continuous: the series **S** of § 3.7 is an example to the contrary.

It is interesting to collate these theorems with Theorems 34 and 35. If f is even, and convex in $(0, 2\pi)$, then $a_n \geqq 0$ for $n > 0$, by Theorem 35. If f is also bounded, then it is continuous; a conclusion which follows, without the use of F.s., from the general properties of convex functions. If f is odd, and decreasing in $(0, 2\pi)$, then $b_n \geqq 0$, by Theorem 34, and so $s_n(\theta)$ is uniformly bounded. Of course Theorem 57 shows more.

6.4. Another theorem of Kolmogoroff. If f is L^2, then $s_n \to f$ (L^2) and, by Theorem 2, there is a sub-sequence (s_{n_ν}) which tends to f p.p. We shall now prove that this is true for certain sequences (n_ν) *independent of* f. We deduce this from a more general theorem in which f is not necessarily L^2. We say that a t.s. has a *gap* (n_ν, n'_ν) if a_n and b_n are 0 for $n_\nu < n \leqq n'_\nu$.

Theorem 83. *If the F.s. of f has an infinity of gaps (n_ν, n'_ν) for which $n'_\nu/n_\nu \geqq \lambda > 1$, then $s_{n_\nu} \to f$ for almost all θ.*

For
$$\begin{aligned}(n'_\nu - n_\nu)(s_{n_\nu} - f) &= (s_{n_\nu+1} - f) + \dots + (s_{n'_\nu} - f) \\ &= (n'_\nu + 1)(\sigma_{n'_\nu} - f) - (n_\nu + 1)(\sigma_{n_\nu} - f) = o(n'_\nu)\end{aligned}$$

p.p., since $\sigma_n \to f$ p.p. Since $(\lambda-1)n'_\nu \leqq \lambda(n'_\nu - n_\nu)$, it follows that $s_{n_\nu} - f \to 0$ p.p.

Theorem 84. *If f is L^2 and $n_{\nu+1}/n_\nu \geqq \lambda > 1$, then $s_{n_\nu} \to f$ for almost all θ.*

We may suppose $n_0 = 0$ and $n_{\nu+1} > n_\nu$. We write

$$P_0 = \tfrac{1}{2}a_0, \quad P_\nu = \sum_{n_{\nu-1}+1}^{n_\nu} A_n(\theta) \quad (\nu > 0);$$

and define two t.s. T_0 and T_1 by

$$T_0 = P_0 + P_2 + P_4 + \ldots, \quad T_1 = P_1 + P_3 + P_5 + \ldots,$$

the blocks of terms P_ν being in each case written out at length and separated by blocks with zero coefficients. Since $\Sigma(a_\nu^2 + b_\nu^2) < \infty$, T_0 and T_1 are the F.s. of functions f_0 and f_1 satisfying the conditions of Theorem 83, and $f_0 + f_1 \equiv f$. Also n_ν is, for every ν, the rank of either the beginning or the end of a gap of each of these series. Hence $s_{n_\nu}(f_0) \to f_0$, $s_{n_\nu}(f_1) \to f_1$, and $s_{n_\nu}(f) = s_{n_\nu}(f_0) + s_{n_\nu}(f_1) \to f$ p.p.

6.5. Strong summability of Fourier series. If s_n is the partial sum of a F.s., then Fejér's theorem and its extensions show that in certain circumstances

$$\sum_0^n (s_m - c) = o(n); \tag{6.5.1}$$

and it might be supposed that the primary reason for this is *cancellation* between the various $s_m - c$. Our next theorem indicates that this is not so, and that (6.5.1) is usually true because most of the $s_m - c$ are themselves small.

Theorem 85. *If f is real, and L^2 near $t = \theta$; and*

$$\int_0^t \{\phi(\theta, u) - c\}^2 \, du = o(t); \tag{6.5.2}$$

then

$$\sum_0^n \{s_m(\theta) - c\}^2 = o(n), \tag{6.5.3}$$

$$\sum_0^n |s_m(\theta) - c| = o(n). \tag{6.5.4}$$

In particular these equations are true, with $c = \phi(+0)$, whenever $\phi(+0)$ exists, and with $c = f(\theta)$ at a point of continuity. If f is L^2 in $(-\pi, \pi)$, then they are true, with $c = f(\theta)$, for almost all θ.

We need only consider (6.5.3), since (6.5.4) follows from it by Cauchy's inequality. We are concerned (except in the last clause of the theorem) with a local phenomenon, and we may therefore suppose, after Theorem 52, that f is L^2 in $(-\pi, \pi)$. Then

$$s_n - c = \frac{2}{\pi}\int_0^\pi g_c(t)\frac{\sin nt}{t}\,dt + o(1) = \frac{2}{\pi}u_n + o(1),$$

and we have to show that

$$F(t) = \int_0^t g_c^2(u)\,du = o(t) \tag{6.5.5}$$

implies

$$S_n = \sum_1^n u_m^2 = o(n). \tag{6.5.6}$$

We write $\quad U_n(t) = \sum_1^n u_m \sin mt, \quad \Omega_n = \sum_1^n |u_m|.$

Then

$$|U_n| \leqq \Omega_n, \quad |U_n| \leqq nt\Omega_n, \quad \Omega_n^2 \leqq nS_n, \quad \int_0^\pi U_n^2\,dt = \tfrac{1}{2}\pi S_n.$$

Also $\quad S_n = \int_0^\pi \frac{gU_n}{t}\,dt = \left(\int_0^{\pi/n} + \int_{\pi/n}^\pi\right)\frac{gU_n}{t}\,dt = J_1 + J_2,$

say. Here

$$J_1^2 \leqq \int_0^{\pi/n} g^2\,dt \int_0^{\pi/n} \frac{U_n^2}{t^2}\,dt \leqq n^2\Omega_n^2 \cdot \frac{\pi}{n}\int_0^{\pi/n} g^2\,dt = o(\Omega_n^2) = o(nS_n),$$

and $\quad J_2^2 \leqq \int_0^\pi U_n^2\,dt \int_{\pi/n}^\pi \frac{g^2}{t^2}\,dt = \tfrac{1}{2}\pi S_n \int_{\pi/n}^\pi \frac{g^2}{t^2}\,dt.$

The last integral, by (6.5.5), is

$$\frac{F(\pi)}{\pi^2} - \frac{n^2}{\pi^2}F\left(\frac{\pi}{n}\right) + 2\int_{\pi/n}^\pi \frac{F}{t^3}\,dt \leqq O(1) + \int_{\pi/n}^\pi o\left(\frac{1}{t^2}\right)dt = o(n);$$

and so $J_2^2 = o(nS_n)$. It follows that

$$S_n^2 \leqq 2(J_1^2 + J_2^2) = o(nS_n),$$

and this is (6.5.6).

It remains to show that (6.5.2) is satisfied p.p. with $c = f(\theta)$; and this will be so if

$$I = \int_0^t \{f(\theta+u) - f(\theta)\}^2\,du = o(t)$$

p.p. But

$$\{f(\theta+u) - f(\theta)\}^2 = f^2(\theta+u) - f^2(\theta) - 2f(\theta)\{f(\theta+u) - f(\theta)\}.$$

Hence

$$I \leqq \int_0^t |f^2(\theta+u)-f^2(\theta)|\,du+2\,|f(\theta)|\int_0^t |f(\theta+u)-f(\theta)|\,du,$$

and each term is $o(t)$ p.p. because f is L^2.

It is usual to express (6.5.4) by saying that $T(f)$ is *strongly summable* $(C, 1)$ to c.

6.6. Another method of summation. Suppose that $\phi(x)$ and its first two derivatives are continuous for $x \geqq 0$; that

$$(6.6.1) \qquad \mu_m = O(m^{-1});$$

and that

$$\Delta_n(m) = \phi(n\mu_m)-\phi\{(n+1)\,\mu_m\}, \quad \Delta_n^2(m) = \Delta_n(m)-\Delta_{n+1}(m).$$

Then it may be verified at once, by the mean value theorem, that

$$(6.6.2) \qquad \Delta_n(m) = O(m^{-1}), \quad \Delta_n^2(m) = O(m^{-2}),$$

uniformly for $n < m$.

Suppose further that Σu_n is any infinite series; that s_n and σ_n are its partial sum and $(C, 1)$ mean of rank n, so that

$$(n+1)\,\sigma_n = s_0+s_1+\ldots+s_n;$$

and that

$$(6.6.3) \qquad t_m = \sum_{n=0}^{m} u_n\,\phi(n\mu_m).$$

Since

$$(6.6.4) \qquad t_m = \sum_{n=0}^{m-1} s_n\,\Delta_n(m)+s_m\,\phi(m\mu_m),$$

t_m is a transform of s_n in the sense of § 5.2.

Theorem 86. (i) *If* Σu_n *converges to* s, *then* $t_m \to s\phi(0)$.

(ii) *If* Σu_n *is summable* $(C, 1)$ *to* s, *then*

$$(6.6.5) \qquad T_m = t_m-(s_m-s)\,\phi(m\mu_m) \to s\phi(0).$$

If also $\mu_m = a/m$ *and* $\phi(a) = 0$, *then* $t_m \to s\phi(0)$.

We need only prove (6.6.5), the other clauses being corollaries.

Transforming (6.6.4) by another partial summation, we obtain

(6.6.6)

$$t_m = \sum_{n=0}^{m-2} (n+1)\,\sigma_n\,\Delta_n^2(m)+m\sigma_{m-1}\,\Delta_{m-1}(m)+s_m\,\phi(m\mu_m).$$

If we take $u_0 = 1$, $u_n = 0$ for $n > 0$, so that $s_n = 1$ and $\sigma_n = 1$ for every n, this becomes

$$(6.6.7)\qquad \phi(0) = \sum_{n=0}^{m-2} (n+1)\,\Delta_n^2(m) + m\Delta_{m-1}(m) + \phi(m\mu_m);$$

and combination of (6.6.6) and (6.6.7) gives

$$(6.6.8)\qquad T_m - s\phi(0) = \sum_n \alpha_{m,n}(\sigma_n - s),$$

where $\alpha_{m,n}$ is $\qquad (n+1)\,\Delta_n^2(m), \quad m\Delta_{m-1}(m), \quad 0$

for $n < m-1$, $n = m-1$ and $n \geqq m$ respectively. It is plain from (6.6.2) that $\alpha_{m,n} \to 0$ when n is fixed and $m \to \infty$. Also

$$(6.6.9)\qquad \sum_n |\,\alpha_{m,n}\,| \leqq H\left(\sum_{n<m-1} \frac{n+1}{m^2} + \frac{m}{m}\right) < 2H.$$

Hence the first two conditions of Theorem 68 are satisfied, and $T_m \to s\phi(0)$ whenever $\sigma_n \to s$.

It should be observed that (6.6.5) remains true in certain cases in which $\sigma_m \to s$ but u_n depends on m as well as on n (so that the hypothesis cannot be stated as one about the summability of Σu_n). If $\sigma_n = \sigma_n(m)$ is uniformly bounded, and tends to s when m and n tend to ∞, so that $|\,\sigma_n - s\,| \leqq \epsilon$ for $m \geqq M(\epsilon)$, $n \geqq N(\epsilon)$, then $T_m \to s\phi(0)$. For then, after (6.6.8) and (6.6.9),

$$|\,T_m - s\phi(0)\,| \leqq \sum_0^N |\,\alpha_{m,n}\,|\,|\,\sigma_n - s\,| + \sum_{N+1}^{\infty} |\,\alpha_{m,n}\,|\,|\,\sigma_n - s\,| \leqq o(1) + 2H\epsilon$$

for $m \geqq M(\epsilon)$.

6.7. Applications. Theorem 86 has various interesting applications to the theory of t.s.

(i) If $u_0 = \frac{1}{2}a_0$, $u_n = A_n(\theta)$ for $n > 0$, so that Σu_n is $T(\theta)$, then

$$\tfrac{1}{2}\{s_m(\theta + \mu_m) + s_m(\theta - \mu_m)\} = \sum_0^m u_n \cos n\mu_m,$$

which is t_m with $\phi(t) = \cos t$. Hence we obtain

Theorem 87. *If $T(\theta)$ is summable $(C, 1)$, to sum s, and $\mu_m = O(m^{-1})$, then*

$$(6.7.1)\qquad \tfrac{1}{2}\{s_m(\theta + \mu_m) + s_m(\theta - \mu_m)\} - \{s_m(\theta) - s\}\cos m\mu_m \to s.$$

In particular

$$(6.7.2)\qquad \frac{1}{2}\left\{s_m\left(\theta + \frac{\rho\pi}{2m}\right) + s_m\left(\theta - \frac{\rho\pi}{2m}\right)\right\} \to s$$

if ρ is an odd integer.

Thus, for example, (6.7.2) is true, with $s = \frac{1}{2}\{f(\theta+0)+f(\theta-0)\}$, when T is the F.s. of a function f for which θ is a point of jump; and it is true p.p. with $s = f(\theta)$.

(ii) The formulae of § 6.6 throw additional light on the 'Gibbs phenomenon' of § 3.12. We observe first that, if f is continuous at θ, it is bounded in an interval round θ. Hence $\sigma_m(\theta+h)$, the $(C, 1)$ mean of the F.s. of f for $t = \theta+h$, is bounded for small h and all m; and, after (5.5.2), it tends to $f(\theta)$ when $h \to 0$ and $m \to \infty$. Also

$$\frac{1}{2}\left\{s_m(\theta+h_m)+s_m\left(\theta+h_m+\frac{\pi}{m}\right)\right\} = \tfrac{1}{2}a_0+\sum_1^m A_n\left(\theta+h_m+\frac{\pi}{2m}\right)\cos\frac{n\pi}{2m}$$

is the t_m of § 6.6, formed from $T(\theta+h_m+\frac{1}{2}\pi m^{-1})$, with $\phi(t) = \cos t$ and $\mu_m = \frac{1}{2}\pi m^{-1}$. Since $\phi(m\mu_m) = 0$, $T_m = t_m$. It now follows from Theorem 86, and the gloss at the end of § 6.6, that *if $f(t)$ is continuous for $t = \theta$, and $h_m \to 0$, then*

$$\frac{1}{2}\left\{s_m(\theta+h_m)+s_m\left(\theta+h_m+\frac{\pi}{m}\right)\right\} \to f(\theta). \tag{6.7.3}$$

We can now prove

Theorem 88. (i) *The Gibbs set at a point of jump of f includes at least the stretch between*

$$c \pm \frac{d}{\pi}\int_0^\pi \frac{\sin t}{t}\,dt = \tfrac{1}{2}\{f(\theta+0)+f(\theta-0)\} \pm \frac{f(\theta+0)-f(\theta-0)}{\pi}\int_0^\pi \frac{\sin t}{t}\,dt.$$

(ii) *The Gibbs set at a point of continuity is a stretch (which may reduce to a point or include the whole real axis) with $f(\theta)$ as its centre.*

The second clause of the theorem follows at once from (6.7.3); for if $s_m(\theta+h_m) \to f(\theta)+\zeta$, and $h'_m = h_m - \pi m^{-1}$, then

$$s_m(\theta+h'_m) \to f(\theta)-\zeta.$$

To prove the first clause, we observe first that, since $\sigma_m(\theta, f) \to c$ and the F.c. of f tend to 0, there is a sub-sequence (m_ν) such that

$$s_{m_\nu}(\theta, f) \to c. \tag{6.7.4}$$

Next, we write (as in §§ 3.12 and 5.5)

$$g(t) = f(t) - \frac{d}{\pi}\mathbf{f}(t-\theta) = f(t)-k(t),$$

and $g(\theta) = c$, so that g is continuous at θ. It follows from (6.7.3) that

$$\chi_m(g) = \frac{1}{2}\left\{s_m(\theta+h_m, g)+s_m\left(\theta+h_m+\frac{\pi}{m}, g\right)\right\} \to c$$

whenever $h_m \to 0$, and so that

$$\chi_m(f) = c+\chi_m(k)+o(1).$$

Taking first $h_m = 0$, and then $h_m = -\pi/m$, and using what we proved in § 3.12 about the F.s. of k, we find that

$$\frac{1}{2}\left\{s_m(\theta,f)+s_m\left(\theta\pm\frac{\pi}{m},f\right)\right\}\to c\pm\frac{d}{2\pi}\int_0^\pi \frac{\sin t}{t}\,dt.$$

It now follows from (6.7.4) that

$$s_{m_\nu}\left(\theta\pm\frac{\pi}{m_\nu},f\right)\to c\pm\frac{d}{\pi}\int_0^\pi \frac{\sin t}{t}\,dt$$

for an appropriate sequence (m_ν); and this completes the proof.

6.8. The existence of the conjugate function. We now prove

Theorem 89. *If f is L, then*

$$\tilde{f}(\theta)=\frac{1}{2\pi}\int_0^\pi \psi(\theta,t)\cot\tfrac{1}{2}t\,dt$$

exists, as a Cauchy integral down to 0, for almost all θ.

We may suppose f real and $a_0 = 0$. Changing our notation slightly, we write

$$\tilde{f}_\eta=\frac{1}{2\pi}\int_\eta^\pi \psi\cot\tfrac{1}{2}t\,dt,\quad \mathrm{f}_\eta=\frac{1}{\pi}\int_\eta^\pi \frac{\psi}{t}\,dt.$$

We have to show that $\tilde{f}_\eta$, or f_η, tends to a limit p.p. when $\eta\to+0$.

(1) Suppose first that f is L^2. Then $\Sigma(a_n^2+b_n^2)<\infty$, so that there is a g of L^2 whose F.s. is $\Sigma B_n(\theta)$; and $\Sigma B_n(\theta)\,r^n$ tends to $g(\theta)$ p.p., by Theorem 75. Also $\Sigma B_n(\theta)\,r^n-\tilde{f}_\eta(\theta)$, with $\eta = \arcsin(1-r)$, tends to 0 p.p., by Theorem 76. Combining these results, we see that $\tilde{f}_\eta(\theta)\to g(\theta)$ p.p.

(2) Passing to the general case, we define sets S and S^*, depending on ϵ, as follows. It will be convenient to write $\bar{E}$ for the complement of any set E in $(0, 2\pi)$, and e for its measure mE.

If
$$F(\theta)=\int_0^\theta f(t)\,dt,$$
then F is continuous and periodic, $F(0) = F(2\pi) = 0$, and $F' \equiv f$. By Egoroff's theorem (ii), there is a closed set S such that

$$(a)\ \ s>2\pi-\epsilon,\quad (b)\ \ |f|\leqq M=M(\epsilon),\quad (c)\ \ \frac{F(\theta+h)-F(\theta)}{h}\to f$$

uniformly in S. It follows from (c) that

$$|F(\theta+h)-F(\theta)|\leqq 2M|h|, \tag{6.8.1}$$

if θ (but not necessarily $\theta+h$) is in S and $|h|\leqq H=H(\epsilon)$.

The set $\bar{S}$ is the sum of a system of non-overlapping open intervals

Δ_n or (ξ_n, η_n), and $\bar{s} = \Sigma\delta_n < \epsilon$. If 0 and 2π are not points of S, we count them as end-points of intervals Δ_n: all other end-points belong to S. If we expand each Δ_n about its centre in the ratio 3 : 1, rejecting any parts outside $(0, 2\pi)$ and combining any overlapping intervals, we obtain a new set of open intervals. The set complementary to these is a closed set S^* included in S, and $s^* > 2\pi - 3\epsilon$. It is sufficient to prove that $\tilde{f}$ exists p.p. in S^*.

We define $P(\theta)$ as the continuous periodic function equal to $F(\theta)$ for 0, 2π, and θ of S, and linear in each Δ_n. Then

$$P(\theta) = \int_0^\theta p(t)\,dt,$$

where $$p = f \text{ (in } S), \quad p = \frac{F(\eta_n) - F(\xi_n)}{\delta_n} \text{ (in } \Delta_n).$$

Since $\Sigma\delta_n < \infty$, $\delta_n < H$ for sufficiently large n, say for $n > N$. If θ is in Δ_n, and $n > N$, then $0 < \theta - \xi_n < H$ and $0 < \xi_n < 2\pi$, so that ξ_n belongs to S. Hence

$$| F(\theta) - F(\xi_n) | \leqq 2M\delta_n, \tag{6.8.2}$$

by (6.8.1). In particular $| F(\eta_n) - F(\xi_n) | \leqq 2M\delta_n$. Thus p is bounded, and so, after (1), $\tilde{p}$ exists p.p. Hence, if we write

$$Q(\theta) = F(\theta) - P(\theta) = \int_0^\theta (f-p)\,dt = \int_0^\theta q\,dt,$$

we need only prove that $\tilde{q}$ exists, or that q_η tends to a limit, p.p. in S^*.

It is plain that Q is continuous and periodic, is 0 for 0, 2π, and θ of S, and that there is a $C = C(\epsilon)$ such that

$$| Q(\theta) | \leqq C \tag{6.8.3}$$

for all θ. Also, if θ is in Δ_n, and $n > N$, then

(6.8.4)

$$| P(\theta) - P(\xi_n) | = \left| \frac{\theta - \xi_n}{\delta_n} \{F(\eta_n) - F(\xi_n)\} \right| \leqq | F(\eta_n) - F(\xi_n) | \leqq 2M\delta_n,$$

(6.8.5)

$$| Q(\theta) | = | Q(\theta) - Q(\xi_n) | \leqq | F(\theta) - F(\xi_n) | + | P(\theta) - P(\xi_n) | \leqq 4M\delta_n,$$

by (6.8.2) and (6.8.4).

(3) Suppose now that θ is in S^*. Then

$$\mathrm{q}_\eta = \frac{1}{\pi}\int_\eta^\pi \frac{q(\theta+t) - q(\theta-t)}{t}\,dt \tag{6.8.6}$$

$$= \frac{1}{\pi}\left[\frac{Q(\theta+t) + Q(\theta-t)}{t}\right]_\eta^\pi + \frac{1}{\pi}\int_\eta^\pi \frac{Q(\theta+t) + Q(\theta-t)}{t^2}\,dt = u_\eta + v_\eta,$$

say. Since $$\frac{Q(\theta\pm\eta)}{\eta}=\frac{Q(\theta\pm\eta)-Q(\theta)}{\eta}\to\pm q(\theta)$$
p.p., u_η tends to a limit p.p.

Finally we must show that v_η tends to a limit p.p., and this will certainly be true if

$$I(\theta)=\int_0^\pi\frac{|Q(\theta\pm t)|}{t^2}\,dt<\infty$$

p.p. in S^*. Taking the upper sign, we have

$$\int_{S^*}I(\theta)\,d\theta=\int_{S^*}d\theta\int_\theta^{\pi+\theta}\frac{|Q(t)|}{(t-\theta)^2}dt\leqq\int_0^{4\pi}|Q(t)|\,dt\int_{S^*}\frac{d\theta}{(t-\theta)^2}.$$

If t is in S, $Q(t)=0$. If t is in Δ_n and θ in S^*, then $|\theta-t|\geqq\delta_n$, and so

$$\int_{S^*}\frac{d\theta}{(t-\theta)^2}\leqq\int_{\delta_n}^\infty\frac{du}{u^2}=\frac{1}{\delta_n};$$

and hence, by (6.8.3) and (6.8.5),

$$\int_{S^*}I(\theta)\,d\theta\leqq2\Sigma\frac{1}{\delta_n}\int_{\Delta_n}|Q|\,dt\leqq2\sum_{n\leqq N}C+8M\sum_{n>N}\delta_n<\infty.$$

This proves the theorem. It follows, as we remarked in § 5.8, that *the c.s. of any F.s. is summable* $(C,1)$, *or* A, *p.p., to* $\bar{f}(\theta)$.

6.9. Convergence factors in Fourier series. We have seen that all F.s., and their c.s., are in various ways 'nearly convergent p.p.' The theorem which follows states one of these ways more precisely.

Theorem 90. *If* $f\sim(a_n,b_n)$, *then*

$$\sum_1^\infty\frac{A_n(\theta)}{\log(n+1)},\quad\sum_1^\infty\frac{B_n(\theta)}{\log(n+1)}$$

are convergent for almost all θ.

If s_n and σ_n are the partial sum and $(C,1)$ mean of Σu_n, then (as in § 6.6)

$$\sum_0^n\lambda_m u_m=\sum_0^{n-2}(m+1)\sigma_m\Delta^2\lambda_m+n\sigma_{n-1}\Delta\lambda_{n-1}+\lambda_n s_n,$$

for any (λ_n). If $\lambda_n=\{\log(n+1)\}^{-1}$ for $n>0$, and $\lambda_0=2\lambda_1$, then $n\Delta\lambda_{n-1}\to0$, $\Delta^2\lambda_m>0$, and $\Sigma(m+1)\Delta^2\lambda_m<\infty$, by (3.6.8); so that $\Sigma\lambda_m u_m$ is convergent whenever σ_m is bounded and $\lambda_n s_n\to0$. If Σu_n is a F.s., or its c.s., then both conditions are satisfied p.p., the first by Theorems 73 and 74, the second by Theorem 64.

6.10. Kuttner's theorem. Theorems 74, 86 and 89 lead simply to an important theorem connecting the convergence of a F.s. and its c.s.

Theorem 91. *If a trigonometrical series $T(\theta)$ is convergent, and its conjugate series $\tilde{T}(\theta)$ summable $(C, 1)$, in a set E of positive measure, then $\tilde{T}(\theta)$ is convergent almost everywhere in E. In particular, if a Fourier series converges in E, then its conjugate series converges almost everywhere in E.*

The second clause is a corollary of the first, since the c.s. of a F.s. is summable $(C, 1)$ p.p.

Next, if $s_m(\theta)$ is the partial sum of $T(\theta)$,

$$\tfrac{1}{2}\{s_m(\theta+\mu_m)-s_m(\theta-\mu_m)\} = \sum_1^m B_n(\theta)\sin n\mu_m$$

is the t_m of § 6.6, formed from $\tilde{T}(\theta)$, with $\phi(t) = \sin t$. Hence, if $\tilde{T}(\theta)$ is summable $(C, 1)$ in E, to sum $\tilde{s}(\theta)$, and $\mu_m = O(m^{-1})$, then

$$(6.10.1)\quad \tfrac{1}{2}\{s_m(\theta+\mu_m)-s_m(\theta-\mu_m)\}-\{\tilde{s}_m(\theta)-\tilde{s}(\theta)\}\sin m\mu_m \to 0$$

in E, by Theorem 86.

Since $T(\theta)$ converges in E, it converges uniformly in a part E^* of E with $mE^* > mE-\epsilon$, by Egoroff's theorem (i). We shall prove that $\tilde{T}(\theta)$ converges p.p. in E^*. If χ is the characteristic function of E^*, then χ is the derivative of its integral for almost all θ of E^*, and we need only consider such θ. For these

$$m\int_{\theta+m^{-1}}^{\theta+2m^{-1}} \chi\,dt \to 2-1 = 1, \quad m\int_{\theta-2m^{-1}}^{\theta-m^{-1}} \chi\,dt \to 1,$$

and therefore

$$I_m(\theta) = \tfrac{1}{2}m\left(\int_{\theta-2m^{-1}}^{\theta-m^{-1}} + \int_{\theta+m^{-1}}^{\theta+2m^{-1}}\right)\chi\,dt > \tfrac{1}{2}$$

for large m. It follows that there are pairs of points $\theta-\mu_m$ and $\theta+\mu_m$, with $m^{-1} \leqq \mu_m \leqq 2m^{-1}$, both lying in E^* (for otherwise I_m could be at most $\tfrac{1}{2}$); and (6.10.1) is true for such θ and μ_m. Also $s_m(\theta+\mu_m)-s_m(\theta-\mu_m) \to 0$, since $T(\theta)$ converges uniformly in E^*; so that

$$\{\tilde{s}_m(\theta)-\tilde{s}(\theta)\}\sin m\mu_m \to 0.$$

But $\sin m\mu_m > \sin 1$; and therefore $\tilde{s}_m(\theta) \to \tilde{s}(\theta)$. This is true p.p. in E^*, and therefore, since ϵ is arbitrary, p.p. in E.

VII. GENERAL TRIGONOMETRICAL SERIES

7.1. Generalities. There are various ways in which we may be led to associate a function f with a t.s. T. The first of these is that which we have regarded as fundamental in the preceding chapters, in which T is the F.s. of f.

A t.s., however, need not be a F.s., and the first problem suggested by a given t.s. is that of deciding whether it is one or not. There may be some other natural way in which we can associate a function with the series. In particular, it may *converge*, and we are thus led to ask whether, if a t.s. converges, it is necessarily the F.s. of its sum. It is plain that the answer cannot be unreservedly affirmative, since the sum need not be integrable. Thus the series (1.3.3) converges for all θ, but to a non-integrable sum.

One special case is particularly fundamental, that in which the sum of the series is zero. If T converges to 0 for all θ, is it necessarily the F.s. of 0? That is to say, is it true that *if T converges to* 0 *for all θ of $\langle -\pi, \pi \rangle$, then $a_n = 0$ and $b_n = 0$ for all n*, so that T vanishes identically? It is plainly the same thing to say that two different t.s. cannot converge to the same sum for all θ; and for this reason the problem is called the problem of 'uniqueness'.

The answers to these questions were given, for the most part, by Riemann, Heine, Cantor, du Bois-Reymond, and de la Vallée-Poussin, and are, broadly, affirmative. If T converges to 0 for all θ, it is identically zero; more generally, if it converges to a function f, subject only to the 'natural' reservations, then it is the F.s. of f. Indeed, as we shall see, more is true, since we can allow exceptional sets in which the hypothesis of convergence fails.

Of course many different t.s. may converge to 0 in any proper part of $(-\pi, \pi)$. Thus if f is L, and 0 in (a, b), where $-\pi < a < b < \pi$, then the F.s. of f converges to 0 (indeed uniformly) in any $\langle \alpha, \beta \rangle$ interior to (a, b).

7.2. The coefficients of a convergent trigonometrical series. If $T(\theta)$ converges, for a special θ, then $A_n(\theta) \to 0$, but a_n and b_n do not necessarily tend to 0; thus $\Sigma \sin(n!\,\theta)$ is convergent when θ is any rational multiple of π. But if $T(\theta)$ converges in a set of positive measure, then its coefficients tend to zero.

Theorem 92. *If $A_n(\theta) \to 0$, and in particular if $T(\theta)$ converges, in a set E of positive measure, then $a_n \to 0$ and $b_n \to 0$.*

Let
$$A_n(\theta) = \rho_n \cos(n\theta + \mu_n),$$
where $\rho_n = \sqrt{(a_n^2 + b_n^2)}$, and suppose that ρ_n does not tend to 0. Then there are a positive δ and a sequence (n_k) such that $\rho_{n_k} > \delta$ and $\cos(n_k\theta + \mu_{n_k}) \to 0$ in E. It follows (since the cosine is uniformly bounded) that
$$\int_{-\pi}^{\pi} \cos^2(n_k\theta + \mu_{n_k})\,\chi(\theta)\,d\theta = \int_E \cos^2(n_k\theta + \mu_{n_k})\,d\theta \to 0,$$
$\chi(\theta)$ being the characteristic function of E. But the integral here is
$$\tfrac{1}{2}\int_{-\pi}^{\pi} \chi\,d\theta + \tfrac{1}{2}\int_{-\pi}^{\pi} \cos 2(n_k\theta + \mu_{n_k})\,\chi\,d\theta = \tfrac{1}{2}mE + \tfrac{1}{2}\int_{-\pi}^{\pi} \cos 2(n_k\theta + \mu_{n_k})\,\chi\,d\theta$$
which tends to $\frac{1}{2}mE > 0$, by Theorem 30. The contradiction shows that $\rho_n \to 0$.

7.3. Riemann's method of summation. The standard method of attack on the problems of § 7.1 was devised by Riemann. If $T(\theta)$ converges in a set of positive measure, then a_n and b_n tend to 0, by Theorem 92, and the series

$$\tfrac{1}{4}a_0\theta^2 - \sum_1^\infty \frac{A_n(\theta)}{n^2} = \tfrac{1}{4}a_0\theta^2 - \Psi(\theta) = \Phi(\theta) \tag{7.3.1}$$

converges, absolutely and uniformly, to a continuous sum. This series is the series obtained from $T(\theta)$ by two formal integrations. Riemann's fundamental idea, which has dominated all later work on these problems, was to argue backwards, by a process of 'generalized second differentiation', from $\Phi(\theta)$ to $T(\theta)$.

We suppose that $T(\theta)$ is any t.s. whose coefficients tend to zero, and define $\Phi(\theta)$ as in (7.3.1). If we write

$$\Delta_h^2 g(\theta) = g(\theta+h) + g(\theta-h) - 2g(\theta) \tag{7.3.2}$$

(for any g), then

$$R_h(\theta) = \frac{\Delta_{2h}^2\Phi(\theta)}{4h^2} = \tfrac{1}{2}a_0 + \sum_1^\infty A_n(\theta)\left(\frac{\sin nh}{nh}\right)^2. \tag{7.3.3}$$

If, in particular, $T(\theta)$ is the F.s. of $f(\theta)$, then

(7.3.4)
$$\frac{1}{2h^2}\int_0^{2h} \phi(\theta,t)\,(2h-t)\,dt = \frac{1}{2h^2}\int_0^{2h} dt \int_0^t \phi(\theta,u)\,du = R_h(\theta),$$

by two term-by-term integrations, permissible after Theorem 44.

The series (7.3.3) reduces formally to $T(\theta)$ for $h = 0$, and this suggests that it should be used to define the sum of $T(\theta)$ when it is divergent. Generally, if

$$u_0+\sum_1^\infty u_n\left(\frac{\sin nh}{nh}\right)^2 \to s \tag{7.3.5}$$

when $h \to 0$, we shall say that Σu_n is *summable* (R) *to sum* s, and write $\Sigma u_n = s\ (R)$.

It is plain from (7.3.4) that $R_h(\theta) \to s$ whenever

$$\frac{1}{t}\int_0^t \phi(\theta, u)\, du \to s,$$

and this is true p.p., with $s = f(\theta)$, after § 4.3. Thus the F.s. of $f(\theta)$ is summable (R), to sum $f(\theta)$, p.p. and in particular at any point of continuity of $f(\theta)$.

The R-method is not a K-method in the sense of § 5.4, since it does not satisfy (5.4.1). It has however a kernel defined by (5.4.2), and (5.4.4) and (7.3.4) show that this kernel is $\pi(2h-\theta)/4h^2$ if $0 \leq \theta \leq 2h$ and 0 if $2h \leq \theta \leq \pi$. It follows that

$$\tfrac{1}{2}+\cos\theta+\cos 2\theta+\ldots = 0 \quad (R) \tag{7.3.6}$$

for $0<\theta<2\pi$. This may be verified directly.

What is most essential for our present purpose, however, is that the R-method is regular in the sense of § 5.2.

Theorem 93. *If Σu_n converges to s, then it is summable (R) to s.*

If $s_n = u_0+u_1+\ldots+u_n$, $s_n \to s$, and we interpret $(\sin nh)/nh$, for $n = 0$, as 1, we have

$$t(h) = \Sigma u_n\left(\frac{\sin nh}{nh}\right)^2 = \Sigma s_n \Delta\left(\frac{\sin nh}{nh}\right)^2 = \Sigma \alpha_{h,n} s_n,$$

say. Here $\alpha_{h,n} \to 0$ when $h \to 0$, for each n. Also $\Sigma\alpha_{h,n} = 1$; and

$$\sum_0^\infty |\alpha_{h,n}| = \sum_0^\infty \left|\int_{nh}^{(n+1)h} \frac{d}{dt}\left(\frac{\sin t}{t}\right)^2 dt\right| \leq \int_0^\infty \left|\frac{d}{dt}\left(\frac{\sin t}{t}\right)^2\right| dt.$$

Hence the R-method satisfies the conditions of Theorem 68 (modified for the continuous parameter), and is regular.

Theorem 94. *If $u_n \to 0$, then*

$$h\sum_1^\infty u_n\left(\frac{\sin nh}{nh}\right)^2 \to 0.$$

In fact the transformation defined by $t(h) = \Sigma\alpha_{h,n} u_n$, where

$$\alpha_{h,0} = \frac{2h}{\pi}, \quad \alpha_{h,n} = \frac{2}{\pi}\frac{\sin^2 nh}{n^2 h} \quad (n>0)$$

is regular. For $\alpha_{h,n} \to 0$ when $h \to 0$; and

$$\sum_{0}^{\infty} |\alpha_{h,n}| = \sum_{0}^{\infty} \alpha_{h,n} = \frac{2h}{\pi} + \frac{1}{\pi h} \sum_{1}^{\infty} \frac{1-\cos 2nh}{n^2} = 1 + \frac{h}{\pi},$$

if $h > 0$, by (3.8.5), so that the sum tends to 1 when $h \to 0$.

It follows from Theorem 94 and (7.3.3) that

$$\Delta_{2h}^2 \Phi(\theta) = o(h) \tag{7.3.7}$$

when $\Phi(\theta)$ is defined by (7.3.1) and a_n and b_n tend to 0.

7.4. The generalized second derivative of a continuous function. Suppose that $g(x)$ is continuous, and that $\Delta_h^2 g(x)$ is defined by (7.3.2). Then we define $\overline{D}_2 g(x)$ and $\underline{D}_2 g(x)$, the *upper and lower generalized second derivatives* of $g(x)$, by

$$\overline{D}_2 g(x) = \overline{\lim_{h \to 0}} \frac{\Delta_h^2 g(x)}{h^2}, \quad \underline{D}_2 g(x) = \underline{\lim_{h \to 0}} \frac{\Delta_h^2 g(x)}{h^2}. \tag{7.4.1}$$

If $\overline{D}_2 g = \underline{D}_2 g$, we denote their common value by $D_2 g$, and call $D_2 g$ the *generalized second derivative* of g. Thus

$$\Delta_h^2(ax^2+bx+c) = 2ah^2, \quad D_2(ax^2+bx+c) = 2a.$$

If $g''(x)$ exists (for a special x), then $D_2 g$ exists and has the same value. For, by Cauchy's form of the mean value theorem,

$$\frac{\Delta_h^2 g(x)}{h^2} = \frac{g'(x+\vartheta h) - g'(x-\vartheta h)}{2\vartheta h} \quad (0 < \vartheta < 1),$$

and the right-hand side tends to $g''(x)$.

It follows from (7.3.3) that $T(\theta) = s$ (R) is equivalent to $D_2 \Phi(\theta) = s$.

If

$$\Delta_h^2 g(x) = o(h), \tag{7.4.2}$$

we say that g is *smooth* at x. In this case

$$\frac{g(x+h)-g(x)}{h} = \frac{g(x-h)-g(x)}{-h} + o(1).$$

Hence, if D^+g, D_+g, D^-g, and D_-g are the four right- and left-hand upper and lower first derivatives of g,* then

$$D^+g = D^-g = \overline{D}g, \quad D_+g = D_-g = \underline{D}g,$$

* The plus and minus signs conveying the distinction of right and left.

$\overline{D}g$ and $\underline{D}g$ being the upper and lower derivatives of g and the distinction of right and left being irrelevant.

It follows from (7.3.7) that *if $T(\theta)$ is a t.s. whose coefficients tend to* 0, *then $\Phi(\theta)$ is smooth everywhere.*

7.5. A theorem about convex functions. If g is convex, then $\Delta_h^2 g \geqq 0$ and so $\underline{D}_2 g \geqq 0$. On the other hand, if g has a non-negative second derivative, then g is convex. The next theorem is a generalization of this familiar result, and is fundamental for all that follows.

Theorem 95. *If g is continuous and smooth in* (a, b), *and*

$$\overline{D}_2 g \geqq 0 \tag{7.5.1}$$

in (a, b), *except perhaps in an enumerable set E, then g is convex. If $\underline{D}_2 g \leqq 0$ (except in E), then g is concave. If $D_2 g = 0$ (except in E), then g is linear. If there is no exceptional set, then the condition of smoothness may be omitted.*

It will be sufficient to prove the first clause, concerning convexity. The second will then follow by changing the sign of g; and the third is an obvious corollary. Finally, the condition of smoothness will be used only in dealing with E.

(i) We may suppose that

$$\overline{D}_2 g > 0 \tag{7.5.2}$$

(except in E). For suppose that the theorem has been proved on this stronger hypothesis, that g satisfies (7.5.1), and that

$$g_n = g + \tfrac{1}{2} n^{-1} x^2.$$

Then $\overline{D}_2 g_n = \overline{D}_2 g + n^{-1} > 0$. Hence (if the theorem has been proved on the stronger hypothesis) g_n is convex for every n; and therefore $g = \lim g_n$ is convex.

(ii) Suppose that g satisfies (7.5.2), except in E, but is not convex. Then we shall prove that E is non-enumerable, and the contradiction will prove the theorem.

Since g is not convex, there is an interval $\langle \alpha, \beta \rangle$, included in (a, b), in which

$$d(x) = g(x) - \frac{(\beta - x)\, g(\alpha) + (x - \alpha)\, g(\beta)}{\beta - \alpha} = g(x) - mx - n$$

is sometimes positive; and the same is true of

$$d_\mu(x) = g(x) - \mu x - n$$

for any μ sufficiently near to m. The maximum of $d(x)$ in $\langle\alpha,\beta\rangle$ is positive, and is assumed for certain values of x in (α,β); and this also is true of $d_\mu(x)$. There is therefore an x_μ (giving the maximum furthest to the right) for which $\alpha<x_\mu<\beta$ and

$$d_\mu(x)\leqq d_\mu(x_\mu)\quad(\alpha\leqq x\leqq\beta),\qquad d_\mu(x)<d_\mu(x_\mu)\quad(x_\mu<x\leqq\beta).$$

Since $d_\mu(x)\leqq d_\mu(x_\mu)$ for x near x_μ,

$$\bar{D}_2 g(x_\mu)=\bar{D}_2 d_\mu(x_\mu)\leqq 0;$$

and, since this contradicts (7.5.2), *x_μ is a point of E.*

Again, since x_μ gives a maximum of d_μ,

$$\frac{d_\mu(x_\mu+h)-d_\mu(x_\mu)}{h}\leqq 0,\qquad \frac{d_\mu(x_\mu-h)-d_\mu(x_\mu)}{-h}\geqq 0,$$

for small positive h. Hence $D^+d_\mu(x_\mu)\leqq 0$ and $D_-d_\mu(x_\mu)\geqq 0$. Also g, and therefore d_μ, is smooth; and hence

$$\bar{D}\,d_\mu(x_\mu)\leqq 0,\quad \underline{D}\,d_\mu(x_\mu)\geqq 0.*$$

But $\underline{D}\,d_\mu(x_\mu)\leqq\bar{D}\,d_\mu(x_\mu)$, and therefore each of these numbers is 0. That is to say, $d'(x_\mu)=0$ and so $g'(x_\mu)=\mu$. It follows that *there is a different point x_μ of E corresponding to each value of μ near m*, and therefore that *E is non-enumerable.*

A simple corollary, which we shall also use, is

Theorem 96. *If g is continuous and smooth, and $\bar{D}_2(g)\geqq c$ in (a,b), except perhaps in an enumerable set E, then*

$$h^{-2}\Delta_h^2 g\geqq c \tag{7.5.3}$$

for $a<x-h<x+h<b$; and if $\underline{D}_2 g\leqq c$, then $h^{-2}\Delta_h^2 g\leqq c$.

Take, for example, the first set of hypotheses, and let $p=g-\frac{1}{2}cx^2$. Then $\bar{D}_2 p\geqq 0$, and therefore, by Theorem 95, p is convex. That is to say, $\Delta_h^2 p\geqq 0$, and this is (7.5.3). The hypothesis of smoothness may again be discarded if there is no exceptional set.

7.6. The theorems of Cantor and du Bois-Reymond. We now apply Theorems 93–96 to t.s., and it will be convenient to begin with a general remark applying to all our results. We are interested primarily in *convergent* series, and state our theorems for them, but we shall never use the full force of the hypothesis of convergence. All our theorems will be true for all series which (*a*) have coefficients tending to 0, and (*b*) are summable ($\dot{R}$), apart

* See the penultimate sentence of § 7.4.

from the exceptional set. These properties follow from the convergence of the series by Theorems 92 and 93, and are the only consequences of the convergence which we use.

Theorem 97. *If two trigonometrical series converge to the same sum, except in an enumerable set E, then they are identical.*

It is plainly sufficient to prove that, if $T(\theta)$ converges to 0, except in E, then $a_n = 0$ and $b_n = 0$. We define $\Phi(\theta) = \frac{1}{4}a_0\theta^2 - \Psi(\theta)$ as in (7.3.1). Since $T(\theta)$ converges p.p., a_n and b_n tend to 0; hence $\Psi(\theta)$ converges uniformly, and $\Phi(\theta)$ is continuous and smooth.* Since $T(\theta)$ converges to 0, except in E, it is summable (R) to 0, except in E; and therefore $D_2\Phi = 0$, except in E. Hence $\Phi(\theta)$ is a linear function $A\theta - B$. Since $\Psi(\theta)$ is bounded, $a_0 = 0$; and therefore $\Phi(\theta)$ is bounded, $A = 0$, and $\Psi(\theta) = B$. Finally, $\Psi(\theta)$, being uniformly convergent, is the F.s. of its sum B; and therefore $B = 0$ and a_n and b_n are 0 for all n.

Theorem 98. *If $T(\theta)$ converges, except in an enumerable set E, to a bounded function $f(\theta)$, then it is the Fourier series of $f(\theta)$.*

Since f is measurable and bounded, it is integrable. If M is the upper bound of $|f|$ (ignoring points of E), then $|D_2\Phi| \leqq$ M, except in E, and $|h^{-2}\Delta_h^2\Phi| \leqq$ M, by Theorem 96. Now

(7.6.1)
$$a_n\left(\frac{\sin nh}{nh}\right)^2 = \frac{1}{\pi}\int_{-\pi}^{\pi} R_h(\theta)\cos n\theta\,d\theta = \frac{1}{\pi}\int_{-\pi}^{\pi} \frac{\Delta_{2h}^2\Phi(\theta)}{4h^2}\cos n\theta\,d\theta.$$

The integrand is uniformly bounded, and tends to $f(\theta)\cos n\theta$, when $h \to 0$, p.p. Hence, making $h \to 0$, we obtain

$$a_n = \frac{1}{\pi}\int_{-\pi}^{\pi} f(\theta)\cos n\theta\,d\theta.$$

Similarly b_n is the Fourier sine-coefficient of f. The theorem is equivalent to Theorem 97 when $f = 0$.

We have remarked already that the conclusions of Theorems 97 and 98 hold for all $T(\theta)$ whose coefficients tend to 0 and which are summable (R), except in E. When there is no exceptional set, they can be generalized still further: we can replace the hypothesis that a_n and b_n tend to 0 by the much more general hypothesis that *$\Sigma n^{-2}A_n(\theta)$ is the F.s. of a continuous function.* For then, denoting this function by $\Psi(\theta)$, and supposing $n > 0$, we have

$$\frac{1}{\pi}\int_{-\pi}^{\pi} \Psi(\theta)\cos n\theta\,d\theta = \frac{a_n}{n^2},$$

$$\frac{1}{\pi}\int_{-\pi}^{\pi} \Psi(\theta + 2h)\cos n\theta\,d\theta = \frac{1}{\pi}\int_{-\pi}^{\pi} \Psi(\theta)\cos n(\theta - 2h)\,d\theta.$$

* Again by § 7.4.

and so

$$\frac{1}{\pi}\int_{-\pi}^{\pi}\frac{\Delta_{2h}^2\Psi}{4h^2}\cos n\theta\,d\theta = \frac{1}{\pi}\int_{-\pi}^{\pi}\Psi\frac{\Delta_{2h}^2\cos n\theta}{4h^2}\,d\theta = \left(\frac{\sin nh}{h}\right)^2\frac{1}{\pi}\int_{-\pi}^{\pi}\Psi\cos n\theta\,d\theta,$$

which is equivalent to (7.6.1). The proof that $T(\theta)$ is the F.s. of f may then be completed as before: the hypothesis that a_n and b_n tend to 0 was needed only to ensure the smoothness of Φ, and is not necessary when there is no exceptional set.

Thus if $\Psi(\theta)$ *is the F.s. of a continuous function, and* $T(\theta)$ *is summable R, for all* θ, *to a bounded* $f(\theta)$, *then* $T(\theta)$ *is the F.s. of* $f(\theta)$. But here we cannot allow even one exceptional point: thus

$$T(\theta) = \tfrac{1}{2}+\cos(\theta-a)+\cos 2(\theta-a)+\dots$$

is summable (R) to 0, except for $\theta = a$, and its $\Psi(\theta)$ is uniformly convergent; but it is not a F.s.

7.7. Unbounded functions: de la Vallée-Poussin's theorem. We have now to extend Theorem 98 to unbounded functions. We shall use de la Vallée-Poussin's majorant and minorant functions. If f is $L(a,b)$, then there are sequences of continuous functions $p_n(x)$ and $P_n(x)$ such that

(i) $$p_n(a) = P_n(a) = 0,$$

(ii) $$p_n(x)\to\int_a^x f(t)\,dt,\quad P_n(x)\to\int_a^x f(t)\,dt$$

uniformly in $\langle a, b\rangle$, and

(iii) $$\overline{D}p_n(x)\leqq f(x)\leqq \underline{D}P_n(x)$$

wherever $f(x)$ is finite.

Theorem 99. *Suppose that* $f(x)$ *is finite, except in an enumerable set* E, *and integrable in* (a,b); *that* $g(x)$ *is continuous and smooth; and that*

$$\underline{D}_2 g(x)\leqq f(x)\leqq \overline{D}_2 g(x)$$

except in E. *Then*

$$g(x)-J(x) = g(x)-\int_a^x dt\int_a^t f(u)\,du \tag{7.7.1}$$

is linear in (a,b). *The condition that* $g(x)$ *should be smooth is unnecessary if there is no exceptional set.*

We take two sequences (p_n) and (P_n) of minorant and majorant functions, and define q_n and Q_n by

$$q_n(x) = \int_a^x p_n(t)\,dt,\quad Q_n(x) = \int_a^x P_n(t)\,dt.$$

Since q_n and Q_n have continuous derivatives, they are themselves continuous and smooth; and they tend uniformly to $J(x)$.

Next,

$$\frac{Q_n(x+h)-2Q_n(x)+Q_n(x-h)}{h^2}=\frac{P_n(x+\vartheta h)-P_n(x-\vartheta h)}{2\vartheta h},$$

where $0<\vartheta<1$, by Cauchy's mean value theorem. Hence

$$\underline{D}_2 Q_n \geqq \underline{D}P_n \geqq f,$$

except in E. And hence, if we write

$$k(x)=J(x)-g(x), \quad K_n(x)=Q_n(x)-g(x), \quad k_n(x)=q_n(x)-g(x),$$

we have $$\overline{D}_2 K_n \geqq \underline{D}_2 Q_n - \underline{D}_2 g \geqq f-f=0.\star$$

Similarly $\underline{D}_2 k_n \leqq 0$. It follows, by Theorem 95, that K_n is convex and k_n concave. But both K_n and k_n tend to k when $n\to\infty$, and therefore k is linear.

We may replace the lower limits a in (7.7.1) by any other values in (a, b).

It is now easy to prove

Theorem 100. *If a trigonometrical series converges, except in an enumerable set E, to a finite and integrable function f, then it is the Fourier series of f.*

For the function Φ of (7.3.1) is continuous and smooth, and $D_2\Phi=f$ except in E, so that we may take $g=\Phi$ in Theorem 99. Hence

$$\Phi(\theta)-\int_0^\theta dt\int_0^t f(u)\,du=\Phi-J$$

is linear. But if $f\sim(\alpha_n,\beta_n)$, then

$$J-\tfrac{1}{4}\alpha_0\theta^2+\sum_1^\infty\frac{\alpha_n\cos n\theta+\beta_n\sin n\theta}{n^2}$$

is linear, so that

$$\tfrac{1}{4}(a_0-\alpha_0)\,\theta^2-\sum_1^\infty\frac{(a_n-\alpha_n)\cos n\theta+(b_n-\beta_n)\sin n\theta}{n^2}$$

is linear. Since the series is uniformly convergent, its sum is bounded; and hence, as in the proof of Theorem 97, $a_n=\alpha_n$ and $b_n=\beta_n$ for all n.

7.8. Generalizations. Our main object in this chapter has been the proof of Theorem 100. We end by a short statement concerning some of its many generalizations.

There are three main directions of generalization (which may naturally be combined). We may generalize (*a*) the concept of integration with which we work, (*b*) the exceptional set E, or (*c*) the hypothesis of convergence. Generalizations of type (*a*) present no special difficulty, granted the necessary definitions, but we are not concerned with them here.

$\star$ $\overline{\lim}\,(u_n-v_n)\geqq\underline{\lim}\,u_n-\underline{\lim}\,v_n$, and similarly for functions of a continuous variable.

Generalizations of type (b) are more difficult. A set E may be said to be a 'set of unicity' if there is no t.s., not identically zero, which converges to zero except in E: thus any enumerable set is a set of unicity. It is not known what are the structural characteristics of such a set. There are certain perfect sets, such as Cantor's and similar sets, which are sets of unicity. But not all null sets are sets of unicity: there are t.s., not identically zero, which converge to 0 p.p.

Some generalizations of type (c) are easy (and are contained, in substance, in what precedes). Thus we may replace the convergence hypothesis of Theorem 100 by 'a_n and b_n tend to 0, and $T(\theta)$ is summable (R) to f, except in E'. Others require only slight developments of our arguments: thus we can prove that $T(\theta)$ is a F.s. whenever its 'upper and lower sum-functions' $\bar{f}$ and $\underline{f}$ are integrable, and finite except in E.

The most interesting questions, however, are those which concern series summable by the methods of Ch. v. These are much more difficult, and there are simple examples which show that we cannot expect such comprehensive answers. Thus

$$\tfrac{1}{2}+\cos\theta+\cos 2\theta+\dots \tag{7.8.1}$$

is summable $(C, 1)$ to 0 for all θ of $\langle -\pi, \pi\rangle$ except $\theta = 0$, and

$$\sin\theta+2\sin 2\theta+3\sin 3\theta+\dots \tag{7.8.2}$$

is summable (A) to 0 for *all* θ.*

We can only mention one or two particularly striking theorems. It was proved by Verblunsky that *if a_n and b_n are $o(n)$, and $T(\theta)$ is summable (A) to 0 for all θ, then $a_n = b_n = 0$ for all n.* In particular, if $T(\theta)$ is summable $(C, 1)$, then a_n and b_n are necessarily $o(n)$, so that *a t.s. summable $(C, 1)$ to 0 for all θ necessarily vanishes identically.* The example (7.8.1) shows that, in this theorem, we cannot admit even one exceptional point. If there are a finite number of exceptional points ξ, then $T(\theta)$ is a finite linear combination of 'singular series' of the type $\tfrac{1}{2}+\Sigma\cos n(\theta-\xi)$. These results have recently been developed much further by Wolf, who has solved this problem for all series summable by any of Cesàro's means; but the full truth concerning summability (A) is still unknown.

* Since $$r\sin\theta+2r^2\sin 2\theta+\dots=\frac{r(1-r^2)\sin\theta}{(1-2r\cos\theta+r^2)^2}\to 0$$
for all θ.

NOTES

We use the following abbreviations for books and periodicals:

H 1, *H* 2: E. W. Hobson, *The theory of functions of a real variable* (vol. 1, ed. 3, Cambridge, 1927; vol. 2, ed. 2, Cambridge, 1926).

KS: S. Kaczmarz and H. Steinhaus, *Theorie der Orthogonalreihen* (Warsaw, 1935).

T: E. C. Titchmarsh, *The theory of functions* (ed. 2, Oxford, 1939).

Z: A. Zygmund, *Trigonometrical series* (Warsaw, 1935).

AM, Acta Math.; *CR*, Comptes rendus (Paris); *FM*, Fundamenta Math.; *JLMS*, Journal London Math. Soc.; *MA*, Math. Annalen; *MZ*, Math. Zeitschrift; *PLMS*, Proc. London Math. Soc. (2).

Other books on the theory of t.s. are H. S. Carslaw, *Introduction to the theory of Fourier's series and integrals* (ed. 3, London, 1930); H. Lebesgue, *Leçons sur les séries trigonométriques* (Paris, 1906); W. Rogosinski, *Fouriersche Reihen* (Leipzig, 1930); L. Tonelli, *Serie trigonometriche* (Bologna, 1928); J. Wolff, *Fourier'sche Reihen* (Groningen, 1931). There are also three important articles of an encyclopaedic character, by H. Burkhardt (II A 12) and E. Hilb and M. Riesz (II C 10) in the *Encykl. d. Math. Wiss.*, and by A. Plessner (I 3, 1325) in Pascal's *Repertorium d. höheren Analysis*. In particular Burkhardt's article contains a mass of information about the early history of the subject.

Chapter I

Nearly all of what we assume in this chapter is in *T*, but we add a few supplementary references. Theorems will not always be found exactly in the form in which we state them, but the reader should have no difficulty in making any adaptations necessary.

§ 1.4. Fubini's theorem is proved in *H* 1, 629; Kestelman, *Modern theories of integration* (Oxford, 1937), 205; Saks, *Theory of the integral* (ed. 2, Warsaw, 1937), 76; de la Vallée-Poussin, *Intégrales de Lebesgue* (ed. 2, Paris, 1934), 54. *T*, 390, proves only a special case. Egoroff's theorem (i) and Fatou's lemma will be found on pp. 339 and 346 of T. Egoroff's theorem (ii) is proved similarly on using Lusin's theorem (*H* 2, 144) and observing that the sets $E_{\eta,H}$ of all x in E for which $|f_h(x)-f(x)|\leqslant\eta$ when $|h|\leqslant H$ are closed in E if $f_h(x)$ and $f(x)$ are continuous in E. For the Stieltjes integral see *H* 1, 538, 662 or Widder, *The Laplace transform* (Princeton, 1941), ch. 1.

§ 1.5. There is a very full discussion of Hölder's and Minkowski's inequalities in Hardy, Littlewood, and Pólya, *Inequalities* (Cambridge, 1934), chs. 2 and 6.

§§ 1.6–1.7. For Th. **1**, *T*, 376; *H* 1, 632 ($p = 1, 2$); *H* 2, 250: for Th. **2**, *T*, 386, 397 (Ex. 17); *H* 2, 254: for Th. **3**, *T*, 397 (Ex. 18); *H* 1, 636, 639 ($p = 1, 2$); *H* 2, 331; *Z*, 17 ($p = 1$), 85.

§§ 1.9–1.11. For all this see *KS* (for orthogonalization, 61).

CHAPTER II

§ 2.1. Th. **11** is 'Bessel's inequality'.

§ 2.2. The 'Riesz-Fischer' theorem was proved independently by E. Fischer and F. Riesz in 1909. There is an interesting historical and critical essay on the theorem by W. H. and G. C. Young in the *Quarterly J. of Math.*, 44 (1913), 49.

§ 2.3. Parseval (in 1799) was the first mathematician to write down formulae expressing sums $\Sigma p_n q_n$ as integrals involving the functions $\Sigma p_n e^{ni\theta}$ and $\Sigma q_n e^{ni\theta}$, and his name is associated generally with theorems of this character.

§ 2.4. Mercer, *Phil. Trans. Roy. Soc.* (A), 211 (1912), 111 (124).

§ 2.5. Th. **18** is proved in *KS*, 200.

§ 2.9. See also § 5.11. There are many direct proofs of Weierstrass's theorem, of which the most elegant is due to S. Bernstein. See *H* 2, 228, 459; Pólya and Szegö, *Aufgaben aus der Analysis* (Berlin, 1925), I, 65, 66, 227, 230; Widder (*l.c.* on § 1.4), 152.

There are important analogues of the Parseval and Riesz-Fischer theorems for functions of a general class L^p. We write

$$\mathbf{S}_r(c) = \left(\sum_{-\infty}^{\infty} |c_n|^r\right)^{1/r}, \quad \mathbf{J}_r(f) = \left(\frac{1}{2\pi}\int_{-\pi}^{\pi} |f|^r d\theta\right)^{1/r},$$

for $r > 1$, and suppose that $1 < p \leqq 2$. Then W. H. Young and Hausdorff (the first for $p = 2, \frac{4}{3}, \frac{6}{5}, \frac{8}{7}, \ldots$, the second for general p) proved the two following theorems: (i) *if f is L^p and $f \sim (c_n)$, then* $\mathbf{S}_{p'}(c) \leqq \mathbf{J}_p(f)$: (ii) *if* $\mathbf{S}_p(c)$ *is finite, then there is an f of $L^{p'}$ with F.c.* c_n, and $\mathbf{J}_{p'}(f) \leqq \mathbf{S}_p(c)$. The restriction $p \leqq 2$ is essential. The theorems were extended to general uniformly bounded o.s. by F. Riesz.

Hardy and Littlewood proved a group of theorems of the same general character, but involving sums of the type $\Sigma |n|^{p-2} |c_n|^p$ and analogous integrals; and these were extended to general o.s. by Paley. For all this see *Z*, ch. 9, and *KS*, ch. 6.

There are many extensions of Parseval's theorem to pairs of functions f and F of associated classes (such as f of L, F of V). The most remarkable is the theorem of M. Riesz, that (2.3.1) *is true when f is L^p and F is $L^{p'}$*. See the note on § 6.8.

CHAPTER III

§ 3.2. The 'Riemann-Lebesgue' theorem was proved by Riemann for 'Riemann-integrable' functions, and extended to functions of L by Lebesgue.

§ 3.3. The theorem in (i) is due to Carathéodory, *MA*, 64 (1907), 95. That in (ii) was proved independently by Dieudonné, *CR*, 192 (1931), 79 and Rogosinski, *Jahresberichte d. Deutschen Math. Ver.*, 40 (1931), *33*. It is still unknown whether $|C_n| \leqq n$ for *all* schlicht functions $F(z) = z + C_2 z^2 + \ldots$.

§ 3.4. If f is analytic and regular on $\langle -\pi, \pi \rangle$, then $c_n = O(e^{-\delta|n|})$ for some positive δ: this condition is also sufficient.

§ 3.5. Th. **41** was proved first (with a quite different example) by F. Riesz, *MZ*, 2 (1918), 312. The proof here is due to Hille and Tamarkine, *Amer. Math. Monthly*, 36 (1929), 255. See also *Z*, 293.

The theorem stated without proof is due to N. Wiener, *Massachusetts J. of Math.*, 3 (1924), 72; see *Z*, 221. It is easy to deduce two interesting theorems concerning the absolute convergence of F.s., due to S. Bernstein and Zygmund respectively: (*a*) *the F.s. of f is absolutely convergent if f is* Lip α, *where* $\alpha > \frac{1}{2}$; (*b*) *it is absolutely convergent if f is V and* Lip α *for some* $\alpha > 0$: here Lip α is the class of functions f for which $|f(\theta+h)-f(\theta)| \leqq H|h|^{\alpha}$. For these see *Z*, ch. 6.

§ 3.8. More generally, a F.s. may be integrated term by term after multiplication by any function of *V*. See W. H. Young, *PLMS*, 9 (1911), 449; Hardy, *Messenger of Math.*, 51 (1922), 186; *H* 2, 581; and *Z*, 91.

§ 3.10. The idea of the proof of Th. **46** is due to Zygmund. Ths. **47** and **48** are due to W. H. Young, *PLMS*, 12 (1913), 41. The proof of Th. **47** may be adapted to show that both C and S are dominatedly convergent when $\Lambda < \infty$: this condition is also necessary for the dominated convergence of either series. In this case C and S also converge strongly with index 1, i.e. $C_n(\theta) \to f(\theta)$ (L) and $S_n(\theta) \to g(\theta)$ (L).

If λ_n is convex, then a necessary and sufficient condition for the strong convergence of C (with index 1) is that $\lambda_n \log n \to 0$: see *Z*, 110. If λ_n is convex and $\Lambda < \infty$, then $f(\theta)\log^+|f(\theta)|$ is *L*.

§ 3.11. Th. **49** is due in part to Chaundy and Jolliffe, *PLMS*, 15 (1916), 214 and in part to Jolliffe, *Proc. Camb. Phil. Soc.*, 19 (1921), 191. In *T* and *Z* the theorem is not stated completely: see *T*, 6 and *Z*, 108.

More generally, $g(\theta) \to \frac{1}{2}\pi A$ is equivalent to $n\lambda_n \to A$, but the proof is more difficult. See Hardy, *PLMS*, 32 (1931), 441; Hardy and Rogosinski, *JLMS*, 18 (1943), 50.

§ 3.12. The properties of the partial sums $s_n(\theta)$ of the special series S have been studied intensively by Gronwall, D. Jackson, Landau and others: Gronwall, *MA*, 72 (1912), 228 should be added to the references in *Z*.

Further theorems concerning the Gibbs phenomenon will be found in §§ 4.5, 5.7 and 6.7. There are graphical discussions in Bromwich, *Infinite series* (ed. 2, Cambridge, 1926), 382 and in Carslaw, I, ch. 9. For the correct numerical value of *G* see Szász, *Duke Math. J.*, 11 (1944), 824.

The phrase 'Gibbs phenomenon' is well established, but Gibbs had been anticipated in the essential idea by Wilbraham (1848) and du Bois-Reymond (1874). Neither Wilbraham nor du Bois-Reymond is quite accurate, and the first general and scientific discussion of the phenomenon was by Bôcher (1906). See Carslaw, *Bull. Amer. Math. Soc.*, 31 (1925), 420.

Chapter IV

§ 4.3. For Lebesgue's theorem quoted here see *H* 1, 637; *T*, 362.

§ 4.6. Gergen, *Quarterly J. of Math.* (Oxford), 1 (1930), 252. The theorem is not stated completely in *Z*. There is a more general form in which (4.6.1) is replaced by

$$\lim_{a\to 0} \overline{\lim_{h\to 0}} \int_{ah}^{\delta} \frac{|\phi(t+h)-\phi(t)|}{t}\,dt = 0:$$

for this see Gergen, *l.c.*, and Pollard, *JLMS*, 2 (1927), 255.

The hypothesis (4.6.5), which implies (4.6.1), is certainly satisfied if $t^{-1}g_c(t)$ is L, and this makes it obvious that Lebesgue's criterion includes Dini's.

There are two other well-known tests, viz.

(i) **de la Vallée-Poussin's test** (V): *if* $\chi(t) = \frac{1}{t}\int_0^t \phi(u)\,du$ *is* $V(0,\pi)$, *then the series converges to* $\chi(+0)$;

(ii) **Young's test** (Y): *if* f *satisfies* l_c, *and the variation of* $ug_c(u)$ *in* $(0,t)$ *is* $O(t)$, *then the series converges to* c.

V includes Dini's test (D) and Jordan's (J), but is included in Lebesgue's (L). Y includes J but not D, and is included in the Pollard-Gergen generalization of L; in the narrower form in which O is replaced by o, it is included in L. In this form it is equivalent to one given by Lebesgue, *MA*, **61** (1905), 251 (énoncé 6, 257). It should also be observed that a test substantially equivalent to V (though stated in terms of the older theories of integration) had been given in 1881 by du Bois-Reymond: see Brodén, *MA*, **52** (1899), 177 (213). For the logical relations between these tests see Hardy, *Messenger of Math.*, **47** (1918), 149; *H* 2, 533; *Z*, 36.

A test of a different type was found recently by Hardy and Littlewood: *the series converges to c if*

$$\frac{1}{t}\int_0^t |g_c|\,du = o\left\{\left(\log\frac{1}{t}\right)^{-1}\right\}$$

and a_n *and* b_n *are* $O(n^{-\delta})$ *for some positive* δ. See Hardy and Littlewood, *JLMS*, **7** (1932), 252 and *Annali d. R. Scuola Norm. Sup. d. Pisa* (2), **3** (1934), 42; Fu Traing Wang, *PLMS*, **47** (1942), 308 and *JLMS*, **17** (1942), 98; *Z*, 34.

Marcinkiewicz, *JLMS*, **10** (1935), 264 has proved that if

$$\frac{1}{t}\int_0^t |f(\theta+u)-f(\theta)|\,du = O\left\{\left(\log\frac{1}{t}\right)^{-1}\right\}$$

for all θ of a set E of positive measure, then $T(f)$ converges p.p. in E.

§ 4.7. Theorem 59 contains the 'Dini-Lipschitz criterion' for uniform convergence. It is a 'best possible' condition in the sense that the o of (4.7.1) cannot be replaced by O: see *Z*, 30, 173. For the inadequacy of the corresponding point criterion see Hardy and Littlewood's second paper quoted above, and *Z*, 174.

§ 4.10. Most of the results of this section are due in substance to W. H. Young, *Münchener Sitzungsberichte*, **41** (1911), 361. 'Dini's test' for the c.s. had been given earlier by Pringsheim.

Th. **63** leads easily to a necessary and sufficient condition for the convergence of $\Sigma n^{-1}a_n$: *the series converges, and*

$$a_0 \log 2 + \sum_1^\infty n^{-1}a_n = \frac{1}{2\pi}\int_0^\pi \cot\tfrac{1}{2}\theta\left\{\int_{-\theta}^{\theta} f(t)\,dt\right\}d\theta,$$

if and only if the last integral exists (*as a Cauchy integral*). This was proved, less simply, by Hardy and Littlewood, *MZ*, **19** (1923), 67 (94).

§ 4.11. Th. **64**: (4.11.1) was proved by Lebesgue, *Ann. de Toulouse* (3), **1** (1909), 25 (116), and (4.11.2) by Lukács, *J. für Math.*, **150** (1920), 107.

§4.12. Th. **65** was first proved by du Bois-Reymond, but the example is due to Schwarz: see *H* 2, 545. There are other very elegant constructions by Faber and Fejér, of which Fejér's is the most familiar: see *H* 2, 541; *T*, 416; *Z*, ch. 8, where further references will be found. Schwarz's example has the advantage that f is defined 'graphically'.

Similar constructions provide examples of continuous functions (*a*) whose F.s. diverge in an enumerable everywhere dense set, (*b*) whose F.s. converge everywhere, but not uniformly in any interval. It is still unknown whether the F.s. of a continuous function (or even one of L^2) can diverge in a set of positive measure.

§4.13. Essentially Lebesgue (*l.c.* on §4.11).

§4.14. The Lebesgue constants have been studied in much detail by Fejér, Gronwall, Hardy, Szegö, and Watson. For references see Hardy, *JLMS*, 17 (1942), 4.

Chapter V

§5.2. We need only the sufficiency of Toeplitz's conditions, which is proved in *Z*, 40. There is a complete proof in Dienes, *The Taylor series* (Oxford, 1931), ch. 12.

§5.3. The $(C, 1)$ method is the simplest of the (C, k) methods defined by Cesàro for integral k and by Chapman and Knopp for all k greater than -1. There are general accounts of these methods in Borel, *Leçons sur les séries divergentes* (ed. 2, Paris, 1928), ch. 6; Dienes, *l.c.*, ch. 12; *H* 2, ch. 1; and Knopp, *Unendliche Reihen* (ed. 2, Berlin, 1924; English translation, Glasgow, 1928), ch. 13. They are regular when $k \geqslant 0$; summability (C, k) implies summability (C, k') for $k' > k$; and summability (C, k), for any k, implies summability (A).

§§5.4–6. It is difficult to give precise references concerning the details of Ths. **70**–**72**, and our treatment has certain points of novelty; but the most essential ideas are Lebesgue's (*l.c.* on §4.11). See *H* 2, ch. 7, and the references in *Z*, ch. 3.

The results of Th. **70** (iv) hold for the $(C, 1)$ and A methods with $H = 1$. There are refinements for functions of L^p. Thus $\mathbf{M}_p\{\sigma_n(\theta)\} \leqq \mathbf{M}_p\{f(\theta)\}$ for $1 \leqq p \leqq \infty$, and $\sigma_n(\theta) \to f(\theta)$ (L^p) for $1 \leqq p < \infty$: see *Z*, ch. 4, where more general classes of functions are considered. Hardy and Littlewood, *AM*, 54 (1930), 81, have shown that, when $1 < p < \infty$, $|\sigma_n(\theta)|^p$ is dominated by a function of L^p. This is not true for $p = 1$ or $p = \infty$. There are extensions of all these results to the general (C, k) methods.

§5.7. The results of Th. **73** are also true, *mutatis mutandis*, for summability (C, k) with $k > 0$. When $k > 1$ the kernel, like that of the A method, satisfies the conditions of Th. **71**, and the method sums the series whenever f satisfies l_c.

We may say that $\phi(t) \to c\,(C, r)$ if

$$\int_0^t (t-u)^{r-1} g_c(u)\, du = o(t^r): \tag{1}$$

thus '$\phi(t) \to c\,(C, 1)$' means that f satisfies l_c. There is a whole scale of theorems connecting such generalized modes of continuity with Cesàro summation of different orders. Thus continuity (C, r) implies summability (C, k) for $k>r$, and summability (C, k) implies continuity (C, r) for $r>k+1$. In particular, *a necessary and sufficient condition that the F.s. should be summable* (C, k) *for some k is that* (1) *should be true for some r*. If f is bounded, or positive, near θ, then all (C, k) methods of positive order are equivalent, and l_c is a necessary and sufficient condition for summability. See Hardy and Littlewood, *MZ*, 19 (1903), 67 and *JLMS*, 1 (1926), 134, and *Z*, ch. 10, for fuller information and references.

§ 5.8. The substance of the theorems concerning the c.s. is due to W. H. Young, *Münchener Sitzungsberichte*, 41 (1911), 361, and Plessner, *Mitteilungen Math. Sem. Giessen*, 10 (1923), 1.

We have not included a general theory of the summation of the c.s., but it may be useful to state the main results here. Two 'conjugate kernels' intervene (as might be expected after the analysis of §§ 5.8 and 5.10), viz.

$$\tilde{K}_m(\theta) = \sum_{n=0}^{\infty} \alpha_{m,n} \tilde{D}_n(\theta) = \sum_{n=0}^{\infty} \alpha_{m,n} \frac{\cos\frac{1}{2}\theta - \cos(n+\frac{1}{2})\theta}{2\sin\frac{1}{2}\theta},$$

$$\tilde{\mathbf{K}}_m(\theta) = \sum_{n=0}^{\infty} \alpha_{m,n} \frac{\cos(n+\frac{1}{2})\theta}{2\sin\frac{1}{2}\theta},$$

and the equation corresponding to (5.4.5) is

$$\int_0^{\pi} \tilde{K}_m(t)\,dt = \sum_1^{\infty} \alpha_{m,n} \log 2n + \gamma + o(1) = l_m + \gamma + o(1).$$

The conditions corresponding to (5.5.1) are

$$\frac{2}{\pi}\int_0^{\pi/m} |\tilde{K}_m(t)|\,dt \leqq H, \qquad \frac{2}{\pi}\int_{\pi/m}^{\pi} |\tilde{\mathbf{K}}_m(t)|\,dt \leqq H;$$

and, if we denote the mth transform of $\tilde{s}_n(\theta)$ by $\tilde{\tau}_m(\theta)$, then the main conclusions corresponding to those of Th. **70** are as follows.

(*a*) If $\psi(+0) = \lim_{t\to 0} \{f(\theta+t) - f(\theta-t)\} = d$ exists, then

$$\tilde{\delta}_m(\theta) = \tilde{\tau}_m(\theta) - r_m \tilde{f}_m(\theta) = \frac{d}{\pi}(l_m - r_m \log 2m) + d\beta + o(1).$$

(*b*) If $f(\theta)$ is continuous in $\langle a, b\rangle$, then $\tilde{\delta}_m(\theta) \to 0$ uniformly in $\langle a, b\rangle$.

(*c*) $\tilde{\delta}_m(\theta) \to 0$ (L).

There are 'L^p' generalisations corresponding to those mentioned in the notes on §§ 5.4–6. As regards β, see the last remark of § 4.8: it is sometimes convenient to use a different value of a.

In the analogues of Ths. **71** and **72**, l_c and L_c must be replaced by $\tilde{l}_d$ and $\tilde{L}_d$, and $\tilde{K}_m$ and $\tilde{\mathbf{K}}_m$ must satisfy conditions similar to those imposed on K_m in those theorems, the first over $(0, \pi/m)$ and the second over $(\pi/m, \pi)$.

In particular $\tilde{\tau}_m(\theta) \to \tilde{f}(\theta)$ p.p. under these conditions: this result naturally depends on Th. **89**. The $(C, 1)$ and (A) methods satisfy the conditions of the theorems corresponding to Ths. **71** and **72** respectively.

§ 5.9. Rogosinski, *Schriften d. Königsberger Gelehrten Ges.*, 1926, Heft 3, proves the generalization of (5.7.5) corresponding to ours of (5.9.4).

§ 5.11. For the 'Tauberian' theorem see *H* 2, 81; *T*, 412; *Z*, 47. When a_n and b_n are $O(n^{-1})$, or real and greater than $-Hn^{-1}$, the convergence problem, both for the F.s. and the c.s., admits a complete solution: the F.s. is convergent to c if and only if f satisfies l_c, and the c.s. is convergent to $\tilde{f}$ if and only if $\tilde{f}$ exists. See Hardy and Littlewood, *JLMS*, 1 (1926), 19.

§ 5.12. Fatou: for further developments and references see *Z*, 256.

Chapter VI

§ 6.2. Kolmogoroff, *FM*, 4 (1923), 324. Later Kolmogoroff constructed a F.s. divergent for *all* θ: see *CR*, 183 (1926), 1327 and *Z*, 175. We use the main idea of the second paper to prove the result of the first.

§ 6.3. Paley, *JLMS*, 7 (1932), 205; *Z*, 265. Szász, *AM*, 61 (1933), 185, has extended the results to the cases $na_n > -H$, $nb_n > -H$: see Hardy and Rogosinski, *l.c.* on § 3.11.

§ 6.4. Kolmogoroff, *FM*, 5 (1924), 96. A t.s. $\Sigma A_{n_\nu}(\theta)$ with $n_{\nu+1}/n_\nu \geqq \lambda > 1$ may be called *lacunary*: Th. **83** shows that such a series, if a F.s., converges p.p. For further results concerning lacunary t.s. see *Z*, 119, 139, 215, 251.

§ 6.5. Th. **85** is the simplest case of a more general theorem: *if f satisfies l_c, and* $\int_0^t |\phi - c|^r du = O(t)$ *for some* $r > 1$, *then* $\sum_0^n |s_m - c|^k = o(n)$ *for all positive* k. See Hardy and Littlewood, *PLMS* (2), 26 (1927), 273 and *FM*, 25 (1935), 162. Marcinkiewicz, *JLMS*, 14 (1939), 162, has shown that *any F.s. is strongly summable p.p.*

For other lines of proof see Fejér, *Proc. Camb. Phil. Soc.*, 34 (1938), 503.

§§ 6.6–7. Rogosinski has made many other applications of the ideas of these sections. See *MA*, 95 (1925), 110; *MZ*, 25 (1926), 132 and 41 (1936), 75. One special result is that

$$s_m\left(\theta + \frac{k\pi}{m}\right) - s_m\left(\theta - \frac{k\pi}{m}\right) \to \frac{2d}{\pi}\int_0^{k\pi} \frac{\sin t}{t}\,dt,$$

if k is an integer and f satisfies $\tilde{L}_d$.

Hardy and Rogosinski, *JLMS*, 18 (1943), 83, give an example of a function whose Gibbs set at θ is not symmetrical about $\frac{1}{2}\{f(\theta+0)+f(\theta-0)\}$.

§ 6.8. Th. **89** is due to Priwaloff, and the usual proof, which uses complex function-theory, to Plessner: see *Z*, 145. Our proof is that of Marcinkiewicz, *FM*, 27 (1936), 38 (55). There is a proof by Besicovitch, *FM*, 4 (1922), 172 and *JLMS*, 1 (1926), 120, independent of the theory of t.s.; but this is decidedly more difficult.

The function $\tilde{f}$ is not necessarily L. It can be shown that if $\tilde{f}$ is L then $\tilde{\mathfrak{T}}$ is the F.s. of $\tilde{f}$, but the proof is rather difficult. M. Riesz proved that if f is L^p, where $1 < p < \infty$, then $\tilde{f}$ also is L^p; and Zygmund that $\tilde{f}$ is L whenever $f(1+\log^+ f)$ is L. For all this see *Z*, ch. 7. Important consequences are that $s_n \to f(L^p)$ and $\tilde{s}_n \to \tilde{f}(L^p)$ when f is L^p and $1 < p < \infty$; that $s_n \to f(L)$ and $\tilde{s}_n \to \tilde{f}(L)$ when $f(1+\log^+ f)$ is L; and that (2.3.1) is true (without absolute convergence) when f is L^p, $1 < p < \infty$, and F is $L^{p'}$.

The last result is true whenever f and $\bar{f}$ are both L. Hardy and Littlewood have shown that in this case $\Sigma n^{-1}(|a_n|+|b_n|)<\infty$. An equivalent result is that the F.s. of f is absolutely convergent if f and $\bar{f}$ are both V. See Z, 139, 157.

§ 6.9. For the F.s., Hardy, *PLMS*, 12 (1913), 365; for the c.s., W. H. Young, *PLMS*, 13 (1914), 13. Kolmogoroff, Seliverstoff, and Plessner have shown that the convergence factor $\{\log(n+1)\}^{-1}$ can be replaced by $\{\log(n+1)\}^{-\frac{1}{2}}$ when f is L^2: see Z, **255**. Later, Littlewood and Paley, *PLMS*, 43 (1937), 105, showed that it can be replaced by $\{\log(n+1)\}^{-1/p}$ when f is L^p and $1<p\leqq 2$; but the proof of this is very difficult.

§ 6.10. Kuttner, *JLMS*, 12 (1935), 131. The proof was simplified by Marcinkiewicz and Zygmund, *FM*, 26 (1936), 1.

Chapter VII

§ 7.1. The theorems of this chapter are to be attributed as follows: **92**, Lebesgue; **93**, **94**, Riemann; **95**, **96**, de la Vallée-Poussin; **97**, Cantor; **98**, du Bois-Reymond; **99**, **100**, de la Vallée-Poussin. Cantor and du Bois-Reymond naturally worked with the older definitions of an integral, and their exceptional sets were less general. Fuller references will be found in H 2, ch. 8, and Z, ch. 11.

§ 7.3. Summability (R) is often called summability $(R, 2)$, summability (R, k) being defined, for $k = 1, 2, 3, \ldots$, by

$$\Sigma u_n \left(\frac{\sin nh}{nh}\right)^k \rightarrow s.$$

This method is regular for $k \geqq 2$ but not for $k = 1$. A good deal has been written recently about the relations between (R, k) and (C, k) summability: for references see Kuttner, *PLMS*, 38 (1935), 273.

It follows from Th. **93** and (7.3.4) that *a necessary condition for the convergence of the F.s. of* $f(t)$, *for* $t = \theta$, *to* c, *is that* $\phi(t) \rightarrow c$ $(C, 2)$.

§ 7.5. de la Vallée-Poussin admits any 'thin' E, i.e. any E which contains no perfect component. The generalization is not wanted here, since the set of points of convergence of a t.s. is 'measurable (B)', and a set which is measurable (B) and thin is necessarily enumerable. See Z, 291.

There is a full discussion of the elementary properties of convex functions in Hardy, Littlewood, and Pólya, *Inequalities*, ch. 3.

§ 7.7. For de la Vallée-Poussin's majorants and minorants see de la Vallée-Poussin, *Cours d'analyse* (ed. 2, Paris, 1909), I, 269, and *Intégrales de Lebesgue*, 68.

§ 7.8. Wolf's work was published in *PLMS*, 45 (1939), 328. The other appropriate references will be found in Z, ch. 11.

A CATALOG OF SELECTED

DOVER BOOKS

IN SCIENCE AND MATHEMATICS

A CATALOG OF SELECTED
DOVER BOOKS
IN SCIENCE AND MATHEMATICS

QUALITATIVE THEORY OF DIFFERENTIAL EQUATIONS, V.V. Nemytskii and V.V. Stepanov. Classic graduate-level text by two prominent Soviet mathematicians covers classical differential equations as well as topological dynamics and ergodic theory. Bibliographies. 523pp. 5⅜ x 8½. 65954-2 Pa. $14.95

MATRICES AND LINEAR ALGEBRA, Hans Schneider and George Phillip Barker. Basic textbook covers theory of matrices and its applications to systems of linear equations and related topics such as determinants, eigenvalues and differential equations. Numerous exercises. 432pp. 5⅜ x 8½. 66014-1 Pa. $12.95

QUANTUM THEORY, David Bohm. This advanced undergraduate-level text presents the quantum theory in terms of qualitative and imaginative concepts, followed by specific applications worked out in mathematical detail. Preface. Index. 655pp. 5⅜ x 8½. 65969-0 Pa. $15.95

ATOMIC PHYSICS (8th edition), Max Born. Nobel laureate's lucid treatment of kinetic theory of gases, elementary particles, nuclear atom, wave-corpuscles, atomic structure and spectral lines, much more. Over 40 appendices, bibliography. 495pp. 5⅜ x 8½. 65984-4 Pa. $13.95

ELECTRONIC STRUCTURE AND THE PROPERTIES OF SOLIDS: The Physics of the Chemical Bond, Walter A. Harrison. Innovative text offers basic understanding of the electronic structure of covalent and ionic solids, simple metals, transition metals and their compounds. Problems. 1980 edition. 582pp. 6⅛ x 9¼. 66021-4 Pa. $19.95

BOUNDARY VALUE PROBLEMS OF HEAT CONDUCTION, M. Necati Özisik. Systematic, comprehensive treatment of modern mathematical methods of solving problems in heat conduction and diffusion. Numerous examples and problems. Selected references. Appendices. 505pp. 5⅜ x 8½. 65990-9 Pa. $12.95

A SHORT HISTORY OF CHEMISTRY (3rd edition), J.R. Partington. Classic exposition explores origins of chemistry, alchemy, early medical chemistry, nature of atmosphere, theory of valency, laws and structure of atomic theory, much more. 428pp. 5⅜ x 8½. (Available in U.S. only) 65977-1 Pa. $12.95

A HISTORY OF ASTRONOMY, A. Pannekoek. Well-balanced, carefully reasoned study covers such topics as Ptolemaic theory, work of Copernicus, Kepler, Newton, Eddington's work on stars, much more. Illustrated. References. 521pp. 5⅜ x 8½. 65994-1 Pa. $15.95

PRINCIPLES OF METEOROLOGICAL ANALYSIS, Walter J. Saucier. Highly respected, abundantly illustrated classic reviews atmospheric variables, hydrostatics, static stability, various analyses (scalar, cross-section, isobaric, isentropic, more). For intermediate meteorology students. 454pp. 6⅛ x 9¼. 65979-8 Pa. $14.95